Control Your **ERP** Destiny

Steven Scott Phillips

Street Smart ERP Publications

Control Your ERP Destiny

Copyright © 2012 by Steven Scott Phillips

Contributors:

Editor: Jamie N. DeMumbrum
Cover and Interior Designer: Emi Ryan
Book Reviewers: Andy Klee, Tom F. Wallace, and R. "Ray" Wang

"SAP" is a registered trademark of SAP AG, Waldorf, Germany.
"JD Edwards" is a registered trademark of Oracle Corporation, Redwood City, California.
"PMI" is a registered trademark of the Project Management Institute, Newtown Square, PA.

Distributed by Lightning Source Inc. (US, UK, AU)

First Printing: May 2012

ISBN-978-0-615-59108-7

Table of Contents

ABOUT THE AUTHOR

Steve Phillips is an ERP professional with over twenty-seven years of implementation experience. His background includes senior management education, software selection, project management, application consulting, process redesign, systems design, testing, training, and post go-live support.

This extensive knowledge is coupled with a rare combination of functional experience in operations management, IT management, and business reengineering. His industry experiences include manufacturing, distribution, business services, and the public sector.

Steve is degreed in Production & Operations Management from The Ohio State University Max Fisher College of Business with advanced studies in Industrial Management from Central Michigan University.

Previously certified by the American Production and Inventory Control Society, he has written articles appearing in APICS publications. He is also a featured contributor for several popular ERP software selection and project management websites.

WHO SHOULD USE THIS BOOK

This book is not academic shelfware. It is designed to facilitate action by providing ERP project management insight that can be successfully applied. When implementing application software it is not only what you do, but how you do it, that can spell the difference between a raging success and an outright failure.

This book will benefit those facing the challenge of implementing ERP or any other complex software package. Also, any organization considering a major upgrade to their existing ERP system will gain valuable information from this book. Specific readers include the project manager, team leaders, business analysts, and application consultants. Many chapters are a must-read for the executive sponsor and IT management.

Partners, practice leaders, and senior managers within ERP software companies and consulting firms that value customer success (more than the billable hours of the moment) can greatly benefit from this book.

Knowledge that enables your clients to understand what project ownership looks like, and to hold up their end of the bargain, is an opportunity largely untapped within the industry. Vendors that embrace the philosophies and strategies contained in this book can help their clients be more successful, enabling the firm to achieve greater success in the future.

HOW TO USE THIS BOOK

First, read this book in its entirety and think about how the information relates specifically to your company, project, and experiences. Next, discuss what was learned with management and the project team, and develop a list of items that may be helpful to your project.

Finally, read each chapter again as the project unfolds. Often it is not until the heat of the battle when it really sinks in. With this approach, management and the team will better understand the decisions at hand, issues and ramifications, possible solutions, or the right questions to ask your consultants.

PREFACE

Shades Off

I must admit, I am a bit of an oddity in the ERP industry. Many consider me a consultant because I once ran with the pack at a Big Five firm. Others like to peg my views as those of a "client practitioner." However, since I no longer use outside software consulting services, how can I be a client? One thing is for sure—I have no software to sell you or any other hidden agenda in writing this book.

Most importantly, I have spent the last twenty-seven years of my life working with real people, in real organizations, and making ERP systems really work. Most of this time was as an "in-house" employee, managing a project and doing the business process design and software details.

Over the course of my career, I have met many wonderful and talented people. I have also observed a lot of crazy stuff. In fact, I have come to an important conclusion I think is worth sharing…

Many conventional ERP wisdoms are simply not working. Those that do are lost in translation or the industry hype.

Let's face it. The track record of the ERP industry is miserable. It has been for a long time and is getting worse. Sadly enough, I am here to tell you the emperor really is not wearing any clothes!

What we need now more than ever is a fresh and different perspective, a voice from the trenches. One based on the organization's success, reality, what works, what does not work, and an understanding of unintended consequences. We must stop the buzzwords, theory, cookie-cutter approaches, or turnkey solutions (that usually exist only in the sales literature).

Also, let us finally acknowledge that organizations are not helpless and discontinue the consulting practices that foster dependency. There is not a consultant in the world that does not advocate the concept of putting ownership in the right place (with their clients). Consultants must stop talking about it and start making it happen.

I hate to say this, but the ERP industry has *institutionalized confusion*. Everyone and their brother is selling ERP software and consulting services. If you have not noticed, there is a feeding frenzy out there, and the meal is the client!

For the typical organization, it is not easy to get a straight answer. One may wonder if there is such a thing as a truly independent source. Unfortunately, in the end, the organization is stuck in the middle and is usually the big loser.

On the other hand, if senior management or others within the company are the problem (as many consultants claim); let's finally get serious about educating them. I know they will listen if consultants can explain the consequences of their actions in terms of the time, money, and misery that lie ahead.

In the meantime, I am not holding my breath. Organizations must take it into their own hands to get educated and to understand how to be more engaged, self-reliant, and have the insight to control their ERP destiny.

This book presents comprehensive strategies to do just that. It can be used within any type of organization and with any ERP implementation methodology. The purpose is to take the discussion to the next level. Given the state of ERP, it is about time someone did.

The definition of ERP success for most companies is very simple: It is about software and business processes that actually satisfy business needs, employees that really understand the system, and implementation costs that are significantly less than "consultant driven" projects.

ACKNOWLEDGEMENTS

Writing this book provides the important opportunity to acknowledge some of the people in my life that made a difference. Beyond those mentioned below, one thing I discovered is everyone has something important to say, whether the CEO of a large corporation or the guy on the receiving dock (making minimum wage). People are people, and this is a good thing.

The lesson is to look for opportunities to learn from everyone. Sort out the nuggets and mix them with your own knowledge and experience. Next, place these newfound revelations safely in your toolbox, because it is a guarantee you will need them later. Specific acknowledgements include:

Betty Phillips (mom) - Upper Sandusky, Ohio. Growing up she made me believe in myself.

Kenneth Phillips (dad) - Upper Sandusky, Ohio. He demonstrated the meaning of integrity, commitment, and humility.

The late **Mae Phillips** (grandma) - Her message: Never give up. Never surrender.

The late **Joseph Orlicky** and **Oliver Wight** - (godfathers of MRP/MRPII/ ERP) - I never met either of them, but their books changed my life.

Sam Powers - (fellow production supervisor at Columbus Show Case Co.) - He took me under his wings when I was a "green" college graduate "new hire" trying to manage production in a plant with five unions.

Brent Gibson - (friend and materials manager at Hill-Rom) - He taught me how to be a knowledge worker: Listen to the employees that perform the work, treat them with respect, learn their jobs, and understand their issues (as well as they do). Next, help develop solutions that work for them and the company.

June Phillips - My wife, best friend, and loving partner. She is the rich earth that enabled me to grow.

"I can feel it! We're about to make a huge breakthrough."

This image is licensed through CartoonStock Ltd., Bath, England.

ERP LESSONS LEARNED

CHAPTER 1
A FALSE SENSE OF SECURITY

Something Needs to Change

Enterprise Resource Planning (ERP) projects are expensive enough, but when not properly managed, can take longer than expected, experience significant cost overruns, and cause business or customer disruption after system go-live. That is, despite all the proven ERP implementation methodologies that have evolved over the years, an alarming percentage of projects continue to have unhappy endings. Could it be some of these methods are not so "proven" after all?

Cite the grim statistics from any independent ERP think tank you want, and the numbers are shockingly similar: Up to 70% of companies that attempt ERP do not realize benefits, go over budget, take longer than expected, or never go-live.

Meantime, consulting costs average 60% or more of total project expenditures. What is wrong with this picture? Unrealistic expectations and poor project estimating are certainly issues, but clearly, something else has gone astray.

A Shift in Philosophy

One would be hard pressed to find any other industry with such a miserable track record, yet it continues with the same old conventional wisdoms. No doubt, we have better ERP software and more implementation tools, but the underlying assumptions about how to run an ERP project have remained unchanged for the past 15 years.

Economist John Keynes once noted, *"Worldly wisdom teaches that it is better for reputation to fail conventionally than to succeed unconventionally."* I believe this speaks volumes to what has been wrong within the ERP industry for many years.

While there are ERP project management practices that do work, many must be revisited, some taken to the next level and others tossed out the window.

Also, for the benefit of all involved, it would be helpful to discontinue the industry jargon and communicate in terms people can understand. Vendors do not seem to realize that most employees participating on a project were yanked from their "normal jobs" to do something they have never done before. If we are to convey meaningful information, we must give them something on which to hang their hats. In other words, "How do I really do this stuff"?

Actually, I sympathize with any practitioner attempting to find actionable information in the mainstream ERP media. The reason is many ERP books, websites, and "free" white papers are published by vendors trying to sell more dependency.

The result is only a fragmented body of knowledge advocating more education, project ownership, knowledge transfer, and self-reliance on the part of the organization. This is the first book to present this body of knowledge as a comprehensive strategy.

While all vendors want their customers to succeed and discuss putting ownership in the right place, in practice, it rarely happens. Of course, organizations must assume some of the responsibility for these failures (which is another reason to write this book). However, I do not see too many vendors rushing to educate their clients in this regard.

Running the Gauntlet

ERP is a multi-billion dollar industry dominated by consultants and software vendors. This is not going to change anytime soon since software and software expertise are the necessities of any automated system. But for a practitioner within industry responsible for a project and a company that must live with the outcomes, the question is: Who solely has your best interest in mind? I can say only one thing: The deck is clearly stacked against you.

First, ERP systems are large in scope, complex to setup, and affect many areas of the business, but this is only half the story. ERP projects have many moving parts that separate them from others involving pure information technology or construction type projects.

ERP implies "big change" and anytime you mix people, culture, software, business processes, consultants, and technology together, the risk increases exponentially. But the real kicker is when the system goes live the ability to run your business depends on getting it right!

For most companies, getting it right is not always easy. Along the way, the organization must sort through a maze of alternative software packages, consultants, and buzzwords, and then, run the gauntlet of implementation to be successful.

During the project, one must manage consultants and contend with software limitations and other broken vendor promises. In addition, we must fix business processes, deal with "people issues" and, often, an organization determined to be its own worst enemy.

ERP Risks Cannot Be Delegated

Successful ERP project management is about understanding and mitigating project risks. When wrestling an 800-pound gorilla, there are inherent risks in everything you do. The ERP pitfalls are many—some obvious, but most not so obvious.

Therefore, ERP risk management can no longer be viewed as a separate line item in the project plan, a one-time brainstorming session, or something we attempt to dump on our consultants. Organizations must understand the risks and potential consequences then make informed project management decisions every day.

For most, the help of consultants is probably required, but hiring an army of software consultants, throwing a few employees at the project, and hoping for the best is no way to manage risk. As ridiculous as this may sound, it happens all the time.

The notion that software consultants or vendors can be the *de facto* owner of your project is not only a naïve concept, but also a very dangerous one. The organization always assumes 95% of the project risks, whether your consultants or your management likes it or not. This is true regardless of the software package selected, the experience of your consultants, and no matter how you structure the contract language.

Vendors Must Be Managed

The intention of this book is not to launch a war against consulting firms and software vendors. I have no personal beef with any of them. I once was an ERP software consultant, know many that still are; and there are plenty of good ones out there. Likewise, there are many good software packages.

Also, even organizations that successfully "go it alone" normally need some level of outside support.

In fact, this book is not about consultants or software vendors at all. It is about business—the business of running a successful ERP project. Managing resources correctly, no matter where they come from, only makes good business sense.

The key is to get educated, assume nothing, and only get the consulting support truly required. Whether you need one or one-hundred consultants, use them for the reasons you brought them in.

This also implies that senior management must get educated, provide internal resources, and invest in their employees. For the investment, we should expect something in return. ERP history has shown that when we expect nothing from the project team, this is exactly what we get. It is too easy to use consultants as a crutch. The problem comes when we are not proactive, do not understand the software details, or do not deliver on our responsibilities. This is when project risk increases substantially.

Do You Call This a Partnership?

It is first important to recognize that software consultants and vendors have some goals that are very different from yours. In my twenty-seven years of implementing ERP, I have concluded there are some things they do not want you to know.

This is not a question of "evil" intent or an industry-wide conspiracy to keep you in the dark, but it is hard to avoid the basic fact: The less you know the more money they make.

Of course, there is nothing wrong with making money. Nevertheless, in an ERP industry where failures are commonplace and consulting costs are out-of-control, it is time to do more than just raise the red flag.

While not every software consultant and vendor will agree with everything I am about to say, for the sake of the project, it is time to be candid and brutally honest. The discussions that follow are long overdue.

The Truth about ERP Software Vendors

For many organizations evaluating software, the euphoric atmosphere during the software demonstrations eventually gives way to the harsh realities of implementation. That is, software vendors are no help in setting realistic expectations and are usually part of the problem.

It must be recognized that the business of peddling ERP software involves huge commissions and the biggest sales pitch on the planet. While software vendors speak of the desire to partner with their clients, how many are eager to tell you all of the things their software does not do? Of course, this is a bit one-sided but you get the picture.

When the prospective buyer attempts to understand what is behind the curtain, most vendors play a *shell game* to conceal software limitations. For example, client questions about the software that result in vendor answers such as *it could, it might, in the future, write a new report, minor software change* or *I will get back with you* usually mean the software does not do it and never will.

When evaluating software, a list of requirements and business scenarios for the vendor to demonstrate are prerequisites. In order to work around these challenges, most vendors agree with what a former US president once said, "It all depends on what your definition of the word *is* is."

While all vendors claim their package contains best practices, *whose* best practice is the multi-million dollar question. Taking a blind leap of faith in this area may be the last leap you ever take.

During the sales process, some vendor practices are downright deceitful, involving power plays or acts of desperation. For example, do not get too excited when your company's terminology is on the demo system menus and screens or some of your data is in the system. These common diversionary tactics do not mean anything when it comes to what the software can do.

Understanding the underlying technologies, integration, and support associated with any package is important. But these are the same reasons some vendors prefer that your IT department not be involved. Instead, they meet with functional managers or end-users who do not know the right technology questions to ask.

In the final analysis, do not be surprised when the vendor attempts to go over the heads of the evaluation team and straight to senior management (particularly if they are losing). This is called *selling solutions to senior management*. All vendors know senior managers are the least qualified people in the organization to be selecting software.

Similarly, any expensive package the vendor is suddenly giving away should raise plenty of eyebrows. We all know what they say about pigs: You can put lipstick on a pig, but it is still a pig. Trust me, the last thing you want to do is buy a *pig in a poke*.

As a member of the team evaluating the software, the vendor might provide free shirts, pens, meals, and maybe even a pass to the national user conference. In return, when a peer asks the vendor a tough question regarding the software, the vendor's hope is that you deflect the question for them.

Software Consultants: The Less You Know, the More Money They Make

Unfortunately, the ERP software consulting industry is not much better than software vendors and it starts at the top of most firms.

For example, within many firms, promotions to partner or senior management positions are based on *selling*, not on successful implementations. This culture can be so intense once the services are sold to the client; the implementation is an afterthought. Ask any experienced software consultant how many times they have been trapped between firm promises versus what can actually be delivered.

Another problem is many firms measure consultant performance in terms of *utilization*. Job performance and bonuses are tied to *billable hours*. It can get to the point that when a consultant is not billing, he or she is not a good consultant even when there is no work to do. This is why some consultants find ways to keep busy (billing) on a project one way or another.

Billable hours drive other agendas that are not always apparent. For example, in some firms a customer with ill-defined project objectives, scope, and an uneducated management team is considered a *gold mine*. The idea of course is to gloss over these "minor" details until after the contract is signed.

Furthermore, during the sales process, what firm does not have the best implementation methods and tools available? However, do not be surprised when their consultants run off and do something entirely different during the project. Maybe the tools are not so great; otherwise, their consultants would use them!

Of course, all consulting firms understand that implementation costs and schedules are always client *hot buttons*. After all, who does not want to install the software quickly and at the least possible cost? Therefore, many firms lowball the quote with the idea they will deliver the bad news later. It is amazing how many organizations fall for the oldest trick in the book!

It does not end here. Once a firm gets its foot in the door, the goal is to *expand* the scope of work, not contain it. Expanding scope is not just the job of firm management. Individual consultants have incentives to get more "experts" in the door or push for software functionality originally out-of-scope.

In addition, do not forget the consultant's promise to transfer software knowledge to the project team. Nine times out of ten, if it is going to happen, you must force the issue. Considering their hourly rates, what incentives do consultants have to transfer software knowledge?

Finally, it can get worse. When a project goes astray, and no one internally understands the software or project details, they become even more dependent on consultants to save the day (possibly the same people that got you into the mess)!

In this case, considering the time and money already invested in the existing consultants, it is not practical to switch firms' midstream. This is when five million dollar projects turn into 15 million dollar disasters.

CHAPTER 2
TAKING PROJECT OWNERSHIP

The Company Is More Than Partially to Blame

In many cases, vendors take the heat (and lawsuits) for ERP disasters when, in fact, it was the client's own doing. Any list of the top five reasons for ERP failure makes it clear that the company is at least partly to blame.

Many like to paint software vendors and consultants as the villains; but even if this is true, guess who bought the software and hired the consultants?

When the organization is not engaged and the project falls hopelessly behind schedule, consultants have no choice but to do it on their own. This not only feeds the consulting cost frenzy, but also senior management is left scratching their heads wondering why the ERP package (used successfully by many in the same industry) does not meet their business needs.

While consultants make a lot of money from uninformed or disengaged clients, no firm wants a black mark on its résumé. All consultants want to meet expectations, but some projects drag on forever because their client is not available, does not resolve business issues, cannot make decisions, or constantly change their minds.

A significant part of the problem is that many have unrealistic expectations of consultants and ERP software. Consultants are only people. They are not all-knowing, do not always make the right decisions, and do not always work on the right things. In addition, when they first walk in the door, they know little, if anything, about your business.

When we allow consultants to run wild, the outcomes are predictable. Do not be surprised to later discover their assumptions, designs, and solutions are totally incorrect. It is never cheap or easy to redesign, reconfigure, and reimplement a botched ERP implementation.

ERP software is another area where many have unrealistic expectations. Here the answer is very simple. It is not surprising that people who write ERP packages do not understand the needs of every company in the entire world!

Software is just software. Companies that fail to do their homework when selecting software are asking for trouble.

Vendors Cannot Own Your Project (Even If You Want Them To)

The first inescapable fact is that vendors will not run their business on the new system, you will. Therefore, the organization ultimately owns the project outcomes.

Second, consultants and software vendors have no direct authority to "make" management or anyone else in the organization do much of anything. Of course, consultants provide expertise, make suggestions, perform the tough system setup tasks, and can insist on many things. But at the end of the day, only the organization can own the project business case, assign the right resources, contain the scope, make decisions, and change business processes. These are a few of the critical success factors only the organization can do and have the biggest impact on project success. No one said it would be easy.

Do Organizations Plan to Fail?

Within the ERP industry, it is common to hear software-consulting firms say: "The system failed because our client did not take ownership of their project." While this is often true, I find this statement a little perplexing. Do senior managers in most companies spend millions of dollars on ERP with the *goal* of failing? Consultants sure make it sound like they do.

Second, where were these same consultants when the disaster was unfolding? Suddenly, in retrospect, they seem to have all the right answers. Maybe they should have explained to their client how to take project ownership from the start.

The truth is, no organization plans to fail—rather, they fail to plan. But many times the issues go deeper than that. ERP is not the type of project most organizations do every day. Unlike more frequently occurring internal projects such as new product development or technology initiatives, not everyone understands the ERP drill, subtle pitfalls, and the consequences of certain decisions.

ERP comes along every ten years or so, and the opportunity to learn and make adjustments through repetition does not exist. In many cases, those heavily involved with the previous project have left the company or want to stay as far away from this one as is humanly possible!

Stop Selling the Organization Short

The good news is that ERP project management deals primarily with management issues that decent managers can understand and do something about. The old saying, *You don't know what you don't know,* certainly applies. However, successful ERP does not require a Ph.D. in computer science or any technology degree, for that matter.

When outside help is required, go get it. But many companies are quick to conclude that they do not have the internal resources to support the implementation. This notion is often advanced by the consulting industry (I wonder why), but it must be challenged from the start.

Contrary to popular ERP wisdom, most companies actually do have some very good employees to assign to the project, if they would just look. Your best employees for ERP are not hard to find. They are the "doers"—employees creditable with management and peers because they have a reputation for getting things done. They are your "go-to people." Every organization has them. Otherwise, how do companies stay in business or get anything accomplished?

In fact, we trust these same employees to perform their daily jobs, rise to the occasion, provide frontline leadership, make decisions, service customers, and help the business grow, adapt, and survive every day. Why is ERP any different?

We can talk all day about the "ideal" skill sets for specific project roles, but rarely does any organization (or consultant) have all of them. Superman does not exist. Nevertheless, many organizations with less than "perfect" employees have been very successful with ERP. What counts are the *collective* knowledge, skills, and experience of a team and their ability to learn—not the heroics of any one individual.

A successful team is made when senior management provides leadership, education, support, and training, and then allows the right employees to do what they are capable of doing (and for which they are accountable). Management must empower employees who are the most knowledgeable about the business and clear the path so they can succeed.

No ERP implementation is ever executed flawlessly. There is always discovery, learning curves, and a few stumbles along the way. However, these same mechanisms foster knowledge, better solutions, and the ability to support the system once installed.

No doubt, staffing the project with the best employees will stretch internal resources. Nevertheless, in most organizations when senior management wants something to happen, they find creative ways to free up the right resources. In fact, this is the first sign of management commitment to anything.

When there are truly not enough internal resources or skills, the answer is simple—go get new employees with the skills. Management does this all the time for other very important initiatives. This approach to project staffing is more easily cost justified than you think. More on this later.

Software Knowledge–The Great Enabler

Rest assured, employees within the company understand the business processes, systems, issues, and improvement opportunities better than anyone does. This should come as no surprise since employees do the work and deal with the problems every day.

So, what if your best employees also had a lot more knowledge of the software? Knowledge of both the business and software details are a powerful force.

On the other hand, a project team that lacks software knowledge increases project risk, implementation costs, and leads to sub-optimization within the business. This is because the team cannot be fully engaged without this knowledge and consultants do not know your business as well as employees do. The result is the software is not leveraged to its potential.

Moreover, an organization that does not know how to "configure" the software to enable required functionality becomes permanently dependent on expensive consultants.

Better Business Solutions (With Fewer Software Consultants)

Conventional ERP wisdom tells us that the more outside experts we have, the better. But history has shown this is usually not the case. It is rarely a good sign when halfway through a project we suddenly need more experts.

Contributing to this mentality is a misguided belief that all software consultants have better ways to run your business. Even if they know the software, many do not understand various industries or how to *redesign* business processes. In this case, their only solution is to throw a new software package into the mix. The notions that all packages are *turnkey solutions* and that all software consultants understand best practices are myths.

This is not necessarily a knock on consultants. For lack of a better choice or at the request of the client, many simply force current processes into software the best they can. Often, the outcome is "square peg meets round hole" and no benefits. It gets worse when the consultant does not know the software.

This highlights the fact that software is not *the* solution. It is only a tool. No solution is complete without changing policies, procedures, workflows, responsibilities, and performance measures. How the software is applied within the business makes all the difference in the world. That is why they called it *application* software!

Real solutions start with an understanding and consensus of the business issues and opportunities. This step is perhaps the toughest part and is where employee knowledge of the business comes in.

Chances are, the best employees have a grasp of the issues and have good ideas on how to make improvements. We must recognize and unleash the collective knowledge, experience, and creativity of employees. It is the job of senior management and the entire project team to enlist the involvement of key employees in order to tap into this vast knowledge base.

This concept is nothing new. Forums and approaches such as *Continuous Improvement, Kaizen, Six-Sigma,* and *Business Reengineering* have been successful for years. While no one is suggesting combining ERP with a full-scale reengineering effort, most practitioners have not connected the dots between these improvement techniques and an ERP implementation.

Of course, software consultants add value in developing solutions. A good consultant can uncover opportunities, make suggestions, and provide guidance based on previous experiences. But a good consultant also knows the client must ultimately own the solution, or it may not be accepted or will not work.

At the same time, too many software consultants can "crowd out" the knowledge and experience within the company that could otherwise play a larger role. Often, the subtle message to employees is "Since we have no confidence in your abilities, we will bring in a bus load of consultants to get the job done." This does not help the cause of employee buy-in and in some cases can lead to outright resistance to any proposed change (no matter how good it is). Here, we are contributing to a *let it fail* mentality within the ranks.

Reduce Software Consulting Cost

In my years of implementing ERP, I have yet to come across any organization that looked forward to paying millions of dollars for software consulting. For smaller organizations planning to spend $100,000, this is no chump change.

No matter how you slice it, software-consulting costs usually represent the *single largest* ERP project expenditure (with, by far, the highest budget risk). Let us explore this opportunity further.

Unlike many other ERP budget items that are somewhat predictable or fixed (such as computer hardware and software), there is a *direct* relationship between consulting cost and the ability of the organization to fulfill its responsibilities (and perhaps do even more).

For example, the cost of the ERP software typically represents about 20% of the total project budget. Many software vendors use the rule of thumb that consulting cost will be roughly two times the cost of the software (or about 40% of the budget). Anyone around ERP long enough understands even this huge number is wishful thinking. After all, what vendor wants to spook a potential buyer away from a software purchase because of the cost to install it?

Depending on the many independent studies one prefers, actual consulting costs can average up to 60% or more of the total budget (or three times the ERP software cost). This is the good news. When ERP projects go bad, they can really go bad. Higher-end ranges for consulting cost are very difficult to come by, but 200% or more of the budget is not surprising.

Therefore, an educated and engaged organization can take a significant bite out of consulting costs. When we take the concept of more self-reliance to the next level (as discussed in the next chapter), up to a 70% reduction in software consulting costs is not out of the question. Again, this is based on my experiences. But at the very least, more ownership by the organization creates a hedge against unexpected events that can otherwise result in major consulting cost overruns.

The New Role of Software Consultants

This leads us to an important conclusion: *It is time to reconsider the traditional roles and responsibilities of software consultants versus those of the organization.*

First, we must get back to what consultants are supposed to be. *Wikipedia* defines a consultant as "one who provides professional or expert advice." Today, most software consultants not only provide advice but also want to do everything!

A software consultant should be a coach, a facilitator, and know when to delegate tasks to their client. A consultant should focus on the transfer of knowledge and perform only the tasks the organization is truly not capable of performing. Given the cost of consultants and the indisputable benefits of client project ownership, is there any other way to run a project?

CHAPTER 3
THE NEXT LEVEL: SELF-RELIANCE

More organizations are taking things into their own hands by developing or acquiring software knowledge previously provided by outside software consultants. In other words, they are performing more than just their traditional project roles, and for good reasons.

Often these same companies once bought into the paternalistic *consultants will take care of us* mindset, were burned, and vowed never to do it the same way again. Others have heard the ERP horror stories such as the company down the street that just went 100% over budget and had little to show for it.

Out of the Closet

The next level is the "other" implementation approach most vendors do not like to mention. When they do, it is called the *go it alone* mentality, casting ominous doubt and fear. Rarely does anyone truly go it alone; but instead, the company uses outside software consultants a lot less and closely manages those they do use.

Many in the ERP industry have successfully practiced this strategy for years, including myself. It is time to drag it out of the shadow and into the light, and recognize it for what it is—a valid approach with a body of knowledge not well-publicized by vendors that have something to lose.

CIO Magazine published the first mainstream article I read acknowledging this topic back in 2002. Written by Christopher Koch (now the Editorial Director at SAP), the title is: "CIOs Take Back Control of Enterprise Projects from Consultants". In this article, on CIO.com, Mr. Koch points out that many CIOs have decided to do most of the ERP project work internally because, given the nature of consulting industry, they believe they have no other choice. My guess is the great majority of organizations would do much more on their own if they knew the right approach.

"It's a Do-It-Yourself Project"

Actually, the concept of putting ownership in the right place has been around from the beginning, but somehow has been lost in the industry hype. Tom F. Wallace an earlier pioneer of MRPII/ERP, states in his book, **ERP: Making It Happen,** the following:

"It's a do-it-yourself project. Successful implementations are done internally. In other words, virtually all of the work involved must be done by the company's own people. The responsibility can't be turned over to outsiders, such as consultants or software suppliers. That's been tried repeatedly, and hasn't worked well at all. Consultants can have a real role in providing expertise but only company people know the company well enough and have the authority to change how things are done. When implementation responsibility is de-coupled from operational responsibility, who can be legitimately accountable for results? If results aren't forthcoming, the implementers can claim the users aren't operating it properly, while the users can say that it wasn't implemented correctly."

Overcome the Fear Mongering

Like anything else, a company doing more of the ERP work internally is not for everyone and will always have its share of naysayers. With that said, it is difficult to change long-standing industry agendas and even senior management mentalities within some companies. Some do not believe it or do not want to believe it. One can call it arrogance, denial, lack of faith, avoiding accountability, or anything else, but many have become so independent of outside consultants that the word "consultant" is not in their vocabulary.

In addition, there are always companies with plenty of money to waste. These companies have several things in common: Management never listens to employees, hires outside experts for everything, and has a history of expensive IT failures. The attitude is *if this does not work, we will buy a different package and use better consultants the next time.* The beat goes on.

I must admit, what I decide to write is not going to make any difference to these folks. If they do not waste their money on ERP, they will waste it on something else! However, in today's economic environment, these types of companies are becoming a rarity.

Finally, there is always the argument that most companies simply do not have enough resources or the right skills to staff the project. While agreed there are exceptions, repeated here are points made in the last chapter. Most organizations have enough internal resources (if they look), have the foundational skills to develop (if they try), or can hire new employees with the skills (more easily cost justified than you think). It is normally not a question of how far an organization can take it, but how far they *want* to take it.

Myth #1: More Risk

Some consultants are quick to conclude that a client doing more on their own is risky business. Interestingly enough, before many of these same consultants were consultants; they were employees within organizations when they developed their software expertise.

Of course, the strategy of developing the software skills internally (vs. hiring new employees with the skills) requires more knowledge transfer and a reasonable period during the project to crest learning curves in order to take more control. During this time, you do lean more on consultants (but less, later). Again, more self-reliance does not imply that consultants disappear off the face of the earth. They should always be viewed as a back up.

Myth #2: Delay System Benefits

The fear mongers also argue that fewer consultants mean the project will take longer, thus delaying system benefits. This is certainly not the case when hiring employees from the outside that already have the right skills.

Developing the resources internally may or may not extend the project schedule, since this depends on the quality of employees assigned and whether or not your consultants are good coaches. But when is a project truly complete? Is it when the system goes live or is it when the software is functioning as desired within the business? These can be two very different timelines.

For example, when the software totally misses the mark, because the organization did not take ownership, the recovery period can take many months or years. During this additional (and unexpected) time to *"get it right"*, the organization realizes no benefits, pays more for consulting, and takes a step backward in productivity.

Invest in Your Own People

As mentioned, developing more software knowledge involves more learning curves and training costs. A CEO might ask, what is the return on investment (ROI)?

While it is clear there is a potential for major savings in software consulting cost, the ROI is anyone's guess. However, ERP is a big investment, and lack of knowledge substantially increases the total cost of ownership and compromises the software investment. This is obvious, even if it is the CEO asking this somewhat rhetorical question.

In addition, outside software consultants face learning curves rarely acknowledged. This learning does not occur overnight and costs you plenty. For example, software consultants must learn the details of their client's business. This includes the business processes, issues, requirements, and data. The project phases *of preliminary analysis, current process analysis*, and *requirements definition*, to a certain extent are about consultants taking the time to understand your business. Something the best employees already know.

Furthermore, there is a misguided belief that experienced consultants face no learning curves with the software. Having spent many years as a software consultant and in-house application expert, I can tell you first-hand that this is not true.

At a minimum, even the best consultants must "test" certain software capabilities (sometimes code for "refresh my memory" or "do something I have never done before").

If for no other reason, consultants face learning curves on every project because the same software tools are flexible to address various business needs. Even minor changes in the software configuration can cause a system to behave very differently. Consulting firms can call this discovery, analysis, testing, or whatever they want, but it is more time and money for the client.

Therefore, no matter how you view it, there are learning curves either way. Why not invest the time and cost of learning in your own employees instead of in consultants that will walk out the door when the project is over?

The Three "In-House" Skills

Ok. This self-reliance theme sounds great in theory, but how do I do it? To get started requires some raw skills in the areas of project management, business analysis, and the ability to learn a software package. The organization does not need an army of employees with these foundational abilities, but just a few.

The remainder of this chapter provides options and justification for developing or acquiring these skills internally.

The Client Is the Project Manager

The organization should drive all aspects of the project including the management of consultants and all other vendors. That is, the client project manager is *the* project manager (PM). After all, it is your project and your system.

Furthermore, a PM from within the company understands the business processes, people, and culture better than any consultant does. When this person also possesses many of the foundational skills (as described in Chapter 12), the advantage is the project manager has instant credibility within the organization. All the above is important when navigating the departmental silos and political waters an ERP project often entails.

Finally, whoever is leading a major change speaks volumes. If nothing else, this is about employee perceptions, and perceptions are 95% of reality. While the PM might need project management support from consultants with the package expertise, this does not mean the consultants should drive the project.

Going Outside for a Project Manager

When a qualified internal candidate cannot be found, smart organizations look externally for a project manager. Here, there are two choices: Hire a new employee or hire an independent, third-party consultant.

Do not fill this role using the project manager of the primary consulting firm involved. First, project management from within the company is a separate role. Second, the project manager must be accountable to the organization and no one else. A PM that is an employee or third party contractor will work in the best interest of the organization, while the project manager from the primary consulting firm has other loyalties.

The minimum skills when hiring a new employee or a third party consultant/contractor as the project manager are:

• Experience managing ERP projects of similar scope and complexity
• Experience within a similar industry
• Ability to quickly learn the organization, adapt, and gain credibility

In addition to these skills, it is always best to hire someone with project management experience with the ERP software selected. This brings knowledge of the implementation steps specific to the package. It also could eliminate the need for PM consulting from the primary firm or reduce this role to that of an account manager (involved only periodically and at a very high-level).

Beyond the project management knowledge relating to the software, the best of all worlds is a PM who also has application consulting experience with the package. This is a tremendous value-add once into the project details.

When it is unlikely you will be evaluating the major ERP packages (Tier 1 or 2), it could be difficult to find a project manager with the package specific knowledge. The typical problem is the limited time after the software is selected and when the implementation phase should begin. When this is a major concern, start the search as early as possible and do not make it dependent on the software.

In fact, there are advantages to bringing in an experienced project manager before selecting ERP software. For example, the PM can join the company early enough to help educate senior management, prepare for the project, and lead the software and consultant selection processes. This can also reduce the cost of consulting services that might otherwise be required during this time and maintains leadership continuity throughout the entire project.

Finally, the basic project phases, deliverables, and key success factors to manage are similar for all ERP packages. Though not ideal, within a reasonable time, any experienced ERP project manager can learn the unique implementation steps and issues associated with any package.

Option 1: Hire a New Employee As the PM

If you must hire a PM from the outside, my first choice would be a permanent employee. This person has a long-term stake in the success of the project since he or she is unlikely to leave the company immediately after go-live.

Admittedly, unlike other project roles, a permanent hire as the PM could be difficult to justify for many companies. This is the situation when the project duration is expected to be short and candidates do not bring "hard" package skills or other abilities required by the company. In other words, once the project is over, what will the person do?

When the opposite is the case, hiring an employee is justified since package-specific skills are always necessary, and there is a role for the person long-term. For example, when the project is over, many ERP project managers eventually become managers within operational functions or the IT department. Other IT roles might include internal application consultant or senior business analyst.

Finally, project management is a standalone discipline. There might be a need for managers for different types of reoccurring projects, some relating to software, and some that are not.

Option 2: Hire an Independent PM Hit Man

Compared to a qualified in-house employee, any type of outside consultant functioning as the internal project manager is never the best-case scenario. When this is the only viable option, hire an independent contractor or firm specializing in the package. In either case, have *one* person in the role (not an entire firm). Remember, too many consultants managing the primary firm is asking for trouble.

Also, many independent contractors have more experience and are less expensive than resources from a traditional firm. The reason is independents need to be better and cheaper since they are the business (or, close to it). They have more of a stake in the outcomes.

In addition, most independents routinely work in these situations and have no ax to grind with consultants from other firms. Typically, their goals are to *get in and get out* and make their money with another successful project under their belt. There is nothing wrong with that.

Missing Link: The Business Analyst

The term *business analysis* is frequently misunderstood, but can spell the difference between success and failure. It is often the missing link between what users *think* they need in the new system versus what the business truly requires.

First, business analysis is not necessarily a single person or role, though it could be. In fact, everyone assigned to the project team should have some business analysis skills. Second, the skill is different from software consulting. A consultant can understand a software package but might have few business analysis skills. This can limited their ability to fully leverage the software capabilities. Also, sometimes we need to find creative ways to work around software limitations.

Business analysis is about developing solutions, many of which have nothing to do with software. The business analyst is collaborative. This person understands the business processes; gathers and interprets user needs; questions practices; identifies problems and opportunities; proposes alternatives; and builds consensus on the solutions that make sense for employees and the organization.

Without a credible knowledge worker (the business analyst), the risk is that system requirements are shaped solely by current operating paradigms, user whims, and the departmental silos. When this occurs, you are automating the mess you already have.

A very good business analyst can analyze any business process and redesign it from scratch (ERP software or no software). Most organizations do not necessarily need this level of business analysis skill for an ERP project. However, if this level is required, do not assume the software consultants have it. The skills and experiences of many consultants have focused primarily on installing a package. There is a difference.

Develop Business Analysis Skills

The raw foundational skills to become a business analyst include being: a quick learner, a logical thinker, detail oriented (yet can see the bigger picture), a problem solver, a team player, and an advocate of positive change (not just any change).

People with these abilities are not standing on every street corner, but we need only a few with the raw talent to develop. It is worth noting that those with business analysis abilities quickly become comfortable working on an ERP project. This type of work is right up their alley.

In addition, depending on the role, each team member needs a different level of business analysis skills. A solid skill set is expected from anyone in an application consulting role. The functional analysts (user representatives) and the project manager should also have these skills. Any business analysis ability the IT group brings to the project is also beneficial.

Additional education and training are helpful to develop business analysts. This includes ERP principles, accepted business practices, problem analysis, and tools such as process mapping or others from continuous improvement disciplines.

Do not limit the perspective in the search for a business analyst. In-house candidates could come from any department. One of the best business analyst I helped develop came straight from the sales department, of all places!

Business analysis skills are somewhat common in the IT department, but where they come from does not matter. Over time, the right person can adapt to any business process, new challenges, and shed any departmental perspective.

Learn How to Configure the Software

Most ERP packages today are "configurable" in the sense the software is flexible to address different business needs without traditional programming (hard coding). The various setup and configuration options, parameters, and switches within the software allow for this flexibility within standard programs.

Knowledge of the configuration settings separates an application analyst from those that only understand how to use the system, or those that write programs. The typical *power user* is not an application consultant without addition training. Also, software development is a different skill, entirely.

The application analyst also understands where the data is stored within the system such as files, fields, and relationships. This knowledge, plus knowledge of the software functionality, is necessary to define programming specifications for software development. An understanding of the data files is also very useful to validate software test results.

Knowledge of the configuration options and data is not technical in nature. Previous experience in IT is not a requirement. The guy from the sales department discussed previously eventually understood every processing option, file, and field in his software area. This is not unusual. In fact, ERP vendors have designed their systems so one does not have be a technical person to configure the application.

Justification for In-House Application Experts

Application expertise is required much longer than most people think. For openers, the average time for an ERP project is roughly 12-36 months. The typical post go-live shake out period is another four to six months. Therefore, just to get the software up and running can average 16 to 42 months.

Beyond the initial start-up, most organizations use their ERP system for another 8 to 12 years. This means the typical organization needs application expertise specific to the package for roughly 9 to 15 years! When viewed from this perspective, developing or hiring in-house application skills makes sense in many ways.

The first major benefit is having someone internal that knows the business, end-users, and can quickly respond to needs. This is important whether the system is hosted internally or outsourced.

When these skills do not exist internally, and software consultants cost too much—this boils down to a failure to leverage the ERP software investment. For example, many companies only use about 60% of the software programs licensed. They might not even know what the other 40% does!

Also, in-house application experts can cost less (in terms of salary and benefits) than hiring outside consultants to perform the initial project and provide support over the life of the system. To understand this better, let's dig deeper into the types of support required once the system is installed. It is important to note most software vendors do not provide this type of support as part of the standard software maintenance agreement.

First, the business is not static. There is always a need to make software configuration changes to satisfy new and changing business requirements for the software already installed.

Second, as employees change jobs and new people are hired; the need for end-user training and support never goes away. Do not count on user word-of-mouth training. When relying on this alone, knowledge of the system and adherence to work procedures slowly erode.

Meanwhile, managers never stop trying to improve the business, and IT never stops trying to replace older systems. Requests to turn-on additional features or install more software modules is a given. In many cases, the organization already owns this software and paying the annual maintenance fees.

When the plan is to make on-going software customizations, as mentioned previously, application knowledge is required. The reason is many software developers need help in analyzing user requests, understanding what the software does out-of-the-box, and the best way to go about making modifications (including avoiding software upgrade difficulties later).

Furthermore, some proposed software modifications are not necessary, since the functionality is satisfied with simple configuration changes. In many cases, only the application analyst has this knowledge.

We are not out of the woods yet. New software releases are available from the vendor about every 12 to 18 months. At some point, most companies want to take advantage of the new functionality in these releases since they already paid for it through software maintenance fees. Implementing a major release is not trivial. The tasks and skills required are similar to the initial install. In other words, you need application people.

More frequent than major releases are interim "patches" or "upgrades" to fix software bugs or incorporate new (or updated) technology associated with the ERP package. If the organization fails to stay current on the technologies, continued software support from the ERP vendor or the technology providers will eventually be at risk. Therefore, an application person must plan and test the updated system before it goes into production.

Contrary to popular belief, the need for internal application expertise does not go away for externally hosted systems. Systems are never really out-of-sight-out-of-mind. When there is no internal analyst to sort out user requirements, make configuration changes, or implement new functionality, it will fall back on the service provider. In this case, the vendor "change orders" may eat you alive.

Option 1: Grow Your Own Application Consultants

Growing your own software consultants starts with selecting the right individuals. Again, the prerequisite skills for an application consultant are not technical in nature, nor require an understanding of traditional programming. The basic skills on which to build are similar to a business analyst.

The other key attributes are intangible. You need someone willing to roll up their sleeves, dig into the software, and learn it (including formal training and coaching from outside consultants). In Chapter 16, I discuss the knowledge transfer steps to develop in-house application consultants.

Option 2: Hire a New Employee with the Application Skills

Given that some organizations do not have the resources to develop application analysts, the solution is to hire permanent employees that already have the knowledge. Unlike developing the role internally, these employees can hit the ground running, virtually eliminating the need for any outside help in this area. A permanent hire is also around long enough to develop more employees into application experts.

The worldwide labor market is saturated with application talent like never before. For many packages, finding the right candidates at the right salary is not hard to do.

Once a project is over, application consultants usually end up in the IT department. This is due to the tight integration of ERP systems and the need for a company-wide perspective when making changes. Allowing power users or employees from various departments to make configuration changes to the system is a disaster waiting to happen. In addition, the application consultant must work closely with the rest of the IT group or software service provider.

CHAPTER 4
ALIGNING SENIOR MANAGEMENT

When embarking on a new ERP system, project expectations can vary widely within the senior management ranks. "Expectation gaps" typically fall into the areas of project benefits, resource commitments, roles, software capabilities, costs, timelines, and training requirements. Worse yet, once the project is underway, there can be major disconnects surrounding the reasons for proceeding with ERP in the first place. When senior management is not on the same page, it is a guarantee that no one else will be either.

Is Senior Management Always the Problem?

In practice, a project manager cannot force upper management to commit to a project. Also, aligning executives and keeping them aligned, in many respects, is like "herding cats." Senior managers, like any one else, have different agendas, priorities, and sometimes play politics.

Nevertheless, when ERP projects disappoint, it is popular (and easy) for consultants or the project manager to simply throw up their hands and cite *lack of senior management commitment* as the reason. However, there appear to be a few holes in this convenient argument:

- In most organizations, do people rise to the senior management level because they are completely incompetent?
- Do most senior managers sign-up for expensive projects like ERP, not intending to support it?
- On most ERP projects, does the executive steering team consider themselves not committed?

The answer for these questions is generally "no," so perhaps a bigger issue is a project management team who lacks a strategy to foster management involvement and commitment.

Managing Executives

Even with the best executive staff, it is foolhardy to assume they automatically understand how to participate. Also, ERP is not the only thing on their minds. They have a business to run that always competes for their attention.

Many times, what is perceived as lack of commitment is really a failure of the consultants and project management team to do their jobs. They are responsible for providing management education and coaching throughout the project. When done correctly, the project is halfway home.

A project manager must manage executives and feel comfortable with "upward delegation" of tasks when necessary. When not fulfilling expectations, the steering team should be reminded of this (of course, in a nice way).

When all fails to get management's attention, there is no magic wand. The last resort is to warn them of the pending disaster and big money ready to go down the tube. Sadly enough, there are projects that come down to this in order to get the steering team's attention.

Uninformed Commitment

It is important to remember successful projects require the right kind of management commitment. *Uninformed* commitment, no matter how well intended, can be as dangerous to a project as no commitment at all.

For example, management should help the team succeed, not just insist on getting it done faster. They also must have realistic expectations regarding the benefits. This is many times a significant problem for the project manager. Education should convey that software will not solve all the problems, and difficult business decisions and changes lie ahead.

What Commitment Looks Like

Senior management commitment to the project is more than just writing the checks and attending steering team meetings. Executives must get out of the executive suite and get involved in planning, supporting, and managing the project.

First, the steering team should be active in developing the business case (project justification) and envisioning the future business state associated with the change. They should define project objectives, approve the scope, and assign the right resources. When developing the project organization, the steering team has project responsibilities, and they must understand how to carry them out successfully.

Their participation in the planning phase creates alignment at the executive level through reconciliation of conflicting views, agendas, and organizational priorities. Not everyone will agree on everything, but by the time the project plan is complete, the goal is to have few, if any, major disconnects at the senior management level. ERP success requires a consistent message coming from the steering team.

Once the project has started, the steering team should participate in project communication and, when requested by the project manager, get directly involved in making decisions or resolving issues.

Executives must *demonstrate* their support through actions, up to and including dealing with employees standing in the way of success. Again, their job is to clear the path for the project team. Without this level of commitment, many employees will not take the project seriously.

Do You Really Need a New ERP System?

This can be the first source of disconnects among executives. Also remember, sometimes the best way to avoid a train wreck is not to get on the train. Executive must be aligned on these basic questions: Is ERP the right solution, and is now the right time to move forward? If there is not consensus on these issues, it will become apparent when the project gets tough.

These questions are not popular among the ERP zealots. In their minds, when you have a hammer in your hand, everything looks like a nail. Nevertheless, not every organization needs a new ERP system, and this must be reconciled.

Make no mistake: There are many reasons to do ERP, including many "no brainers." However, it is not always so obvious for those sitting in the executive suite. The problem is when the ERP bandwagon starts to roll and no one in management is asking the right questions, those with legitimate concerns get steamrolled. It is often not even a fair debate. After all, those pushing for a new system compare the very worst of the current environment to an idealistic ERP concept that, at this stage, is *all things to all people*.

When the business case for ERP is not clear, consider performing a *Business Needs Assessment*. This is different from an Information Needs Assessment. The former term is more appropriate, since information implies a systems assessment, and the analysis should be much more.

In addition, a business needs assessment is different from an ERP Readiness Assessment. The latter suggests ERP is the direction, while a business needs assessment helps determine the directions, and ERP may not be one of them.

A business needs assessment is a high-level analysis of the organization and issues associated with current processes, systems, or anything else that inhibits business performance or achieving the desired state. It may require the help of an independent business consultant, or it may not. It could be as simple as insisting on answers from a group of key managers and employees for the questions below.

Thirteen Questions to Ask before Considering ERP

1. **Would ERP be one of the top priorities, considering other internal projects, initiatives, or external events?** The project must be a focus of management and internal resources.

2. **What is really broken, the software or current business processes?** Many times, so called software limitations are not limitations at all, when attempting to automate bad business practices.

3. **When process deficiencies are the main problem, have we tried to fix the issues without new software?** Ineffective policies, procedures, measurements, and cultural issues typically have nothing to do with software, but much to do with poor business performance.

4. **Will the availability of better information result in better decisions or will it make lousy managers more effective?** Systems never overcome poor management.

5. **Does anyone understand the capabilities within the current software that are not utilized?** Maybe the existing system can be leveraged more. Often the capabilities of a new package look appealing, but many are unaware the current system is capable of doing the same thing.

6. **Is the promise of "new technology" a valid reason to throw out application software that tends to work fine?** Technology is great, if it solves business problems.

7. **Do we have bad software or bad data (garbage in, garbage out)?** Clean up the data and maybe the existing system is very usable.

8. **Is everything about the current software terrible?** With new software, there could be areas where a major step backward in functionality is inevitable. This is especially true if modifications have been made to the current system that are unique to your business and are very useful.

9. **Is the current software really on the brink of being "not supported" by any vendor?** They have said this for years, yet the system keeps running! If there is a problem with support from the primary vendor, what other support options are available? Usually as long as the system is stable, replacement parts are available for hardware, and there are IT resources to support the system, application software can run for many years beyond the official end-of-life date published by the vendor.

10. **Can a few software modifications or enhancements to the current software satisfy 80% of the important needs–for a fraction of the time and cost?** Modifications should be discouraged, but sometimes they make perfect sense.

11. **Can a purchased *bolt-on* package do the trick versus buying an entirely new ERP system?** Bolt-on applications can be the right answer to fulfill very specific needs (when integration with the existing ERP system is not a major issue).

12. **Are the savings associated with new software real or fluff?** If you listen to ERP vendors and do not perform your own objective analysis, it might be fluff.

13. **Have you considered all the implementation and support costs in the return on investment?** Many are not obvious and can greatly increase the cost of ownership.

Reengineering Versus a Package Implementation

In many companies, the underlying issues impeding productivity and performance improvements are much more fundamental than software. In this case, one should consider a new software package only after restructuring how the business operates. The first goal should be to fix the business.

In fact, a basic philosophy of reengineering is to stop using system limitations as an excuse for inaction. Plenty of major improvements are usually possible, without new software.

History has shown attempting to install an ERP package within a business that needs reengineered increases implementation costs, extends the schedule, and compromises the software investment.

The basic issue is that the existence of new software does not necessarily change longstanding paradigms, structural issues, or measurements that drive sub-optimization. If these issues are not at least partially addressed before the ERP project begins, the software side of the project will be in constant conflict with the unaligned organization.

Attempting to install a package and reengineer at the same time is not a good idea either. This is a project that is too large to manage and in a constant state of flux. It is also very inefficient as software consultants sit around waiting on management decisions and the new business model to emerge.

Reengineering and software package implementations are two very different initiatives. The goal of ERP is to *integrate* the information flow across the business, realize *software-enabled* benefits, and take advantage of other *improvement opportunities* presented. Of course, for many the overall benefits are significant.

On the other hand, business reengineering, in its purest sense, is about throwing out many of the current philosophies, business processes, and more or less, starting over. While few take it to this extreme, it is an entirely different focus.

When reengineering occurs *before* purchasing new software, there is a greater chance of defining the "real" software requirements. This is because the requirements are now based on your best practices, not the unaligned organization. At this point, a new system might enable further improvements.

Organize Sooner

When ERP is a go, realize that some of the worst mistakes occur early in the project. The sooner one selects a project manager and organizes the steering team, the better. This ensures that the right people are involved during the critical phases of education and the selection of software and consultants.

When the plan is to hire a project manager from the outside with package experience, assign someone internally as the interim project manager until a package is selected.

Finally, it is always wise to start the steering team education before signing any vendor contracts. This helps management make sound decisions early or perhaps pull the plug on the project before making a large financial commitment.

More than a PowerPoint Presentation

Management education addresses five basic topics:
1) What is ERP?
2) What are the potential benefits?
3) What does it take to get there?
4) What are senior management's responsibilities?
5) What are the lessons learned (success factors)?

When planning education, recognize that it is a journey, not an event. A one-hour presentation on ERP is not going to cut it. It requires much more education—and coaching throughout the project.

There are many resources available to educate executives, including independent ERP education specialists, industry societies, books, and learning from other companies that recently completed an ERP project. Once the project is underway, your software consulting firm should provide additional education, primarily through coaching.

Ignorance Is Bliss (and Convenient)

Many times the project manager must sell the need for senior management education. This issue is if management is not educated, they may not understand why they need to be educated.

Part of the problem is telling executives they lack education with regard to ERP. Instead, use terms such as ERP *management awareness* and *lessons learned.*

For all the seemingly valid arguments for skipping education (discussed next), always remember there are consequences. Take a stand now, or pay the price later when the steering team makes a difficult project even more difficult. The following are typical objections to education raised by senior managers, as well as suggested responses.

1. *I have been through ERP before at another company and understand what we are up against.*

    If this is the case, you may have found an ally in management that can help enlighten others. Nevertheless, at the executive level we need more than one lone voice in the woods. What matters is that all senior managers "get it."

    Even though most ERP projects have similar critical success factors, every organization and project is different in terms of those that rise to the top of the list. Unless someone was heavily involved in more than one project, chances are there is something important to learn.

2. *Since ERP is a computer project, maybe only IT management should get educated.*

    Paraphrasing the late Ollie Wight (the MRPII guru of the 1970s from which ERP evolved): "MRP [ERP] is not a computer system. It is a people system made possible by the computer."

    ERP is not about technology—it is about changes in business processes, policies, responsibilities, and sometimes culture. IT plays an important role, but management must understand and own the business decisions affecting success or failure.

3. *The project manager is the one who needs to know that stuff.*

    Yes, there is a project manager; but given the scope of most ERP projects, project management is not something one can simply delegate to a single individual (if you want to be successful).

At times, it takes the collective brainpower and experience of the entire steering team to bring the project home successfully.

In addition, unless the project manager is the president of the company (not recommended), the PM has no direct authority over most of the employees affected. When push comes to shove (and it always does), the project manager will be left to fend for himself or alternatively seek support from an informed and engaged executive team.

4.   *Education costs too much and takes time.*
There is no need to spend a fortune to educate executives. Education will cost something, but is a small price to pay for a chance at success. In fact, when done correctly, it is the biggest bang for the money spent.

Finally, if management cannot make this time and financial sacrifice for something that will impact the entire business, one can only imagine the level of support they will provide during the implementation.

The Need for Independent Educators

Throughout this book, the need for independent sources for different types of education, training, and consulting is addressed. Senior management education is one of those types.

Independent means the provider is not in any way affiliated with any ERP software vendor or implementation services firm. Avoid so-called educators with an agenda to sell software or consulting services, or have any other reason to sugarcoat the real issues. The reason is this type of education has very little to do with software and nothing to do with a particular package.

The concept of independence usually eliminates the large consulting firms, boutique shops aligned with a particular software package, and large technology companies. No matter what they say, rarely can these vendors be totally objective.

For example, even when the educator or consultant is from a separate business unit within a larger firm that has associations with an ERP package, they might receive pressure from the software side of the house to get their foot in the door. This leaves mostly smaller firms or ERP education specialists as truly independent sources.

Management Education Outline

Part of the value of education is getting the steering team in the same room, at the same time, to hear the same thing. The following is a sample agenda, when planning formal or informal educational activities.

- **ERP Concepts and Principles**
  - Definitions
  - Business Integration
  - Software Modularity
  - Software "Flexibility"

- **Potential Benefits (Do not oversell!)**
  - Single integrated database / "One version of the truth"
  - Industry specific benefits
    - Inventory reduction (as an example)
  - Increase productivity
    - Reduce cycle times
    - Reduce errors
    - Reduce manual processing
  - Eliminate support cost of multiple systems
  - Regulatory / Customer Compliance

- **Implementation Processes (How do we get there?)**
  - Traditional or rapid deployment (pros and cons)
  - Major Project Phases
  - Major Deliverables (within each phase)
  - Typical Project Timelines (12-36 months)
  - Implementation Roll-out Options (pros and cons)

- **Project Organization**
  - Structure / Reporting relationships
  - Role & Responsibilities

- ○ Executive Sponsor
- ○ Steering Team
- ○ Project Manager
- ○ Project Team

- **Critical Success Factors**
  - Senior Management Involvement and Support
  - Manage Scope
  - Redesign Processes
  - Knowledge Transfer
  - Data Accuracy
  - Resource Commitments
  - Accountability
  - Control Software Modifications
  - Stable Technology
  - Best Fit ERP Software
  - Experienced Consultants

- **Implementation Cost Areas**
  - ERP Software
  - Computer Hardware
  - Computer Software
  - Industry Practices Education
  - Software Training
  - Consulting (project management, application, developers, technical)
  - Implementation Tools
  - Ongoing Software Maintenance Fees
  - Implement Upgrades/New Releases

Education on New Operating Strategies

When the project includes some major changes in operating philosophies, additional management education could be necessary. Incorporating a new operating concept is often a greater challenge than applying software to the current ways of doing business.

Within the current environment, at least everyone understands the rules of the game. With an entirely new philosophy, management and employees must grasp new concepts, understand the potential issues, and know how to sustain the changes, once implemented. For example, existing performance measures or management policies may conflict with the new operating practice and, therefore, some of these must change.

Often, management wants the benefits of a best practice but does not understand what it takes to achieve them. This is a big contributor to many ERP project disappointments. Normally, this type of education is not delivered from the typical ERP educator, but from an independent consultant specializing in the industry practice.

Learn the Hard Lessons the Easy Way

A relatively inexpensive, yet effective, way to increase executive awareness is through site visits to other companies that have recently installed ERP. When done correctly these experiences can be a revelation for many, and hammer home topics of the formal education sessions.

Of course, the concept of a site visit is nothing new since they are frequently used for other purposes such as evaluating software or learning best practices. But a site visit for the purpose of senior management education is a good tool no matter what software or practices the host company uses. The primary reason for this type of visit is not software related.

Executive visits should focus on ERP project management and the organizational challenges encountered during the implementation of *any* package. The goal is for executives to walk away with an understanding of what went right with the project, what went wrong, and what should be done differently.

However, before executives sign-off on a team software recommendation, meeting with another company that recently installed the same package can serve the dual purpose of executive education and provide them with a comfort level (by talking to people using the recommended system).

Never send executives to visit a company using an ERP package currently under consideration if the evaluation team has yet to make the final software recommendation. This could taint the selection process.

Site visits can have different objectives, but one common outcome is they often fail to meet expectations. This is usually because expectations were wrong to begin with, or the visit was poorly planned and organized.

For example, regardless of the ERP system the host company is using, do not turn a site visit (for the purposes of executive education) into a software demonstration. This is a common mistake since executives seeing the software function is of very limited value at this stage. Again, this is not the objective of management education, and there will be time to demonstrate the software to executives later (once a package is selected or as part of the final approval).

The only exception is when visiting a company using the same software you plan to purchase and are utilizing operating philosophies you want to implement with the new software. In this situation, there could be value in executives understanding how the software supports the major business change and the management issues to address.

Furthermore, it is usually not necessary to travel the globe to find a company worth a visit. Of course, it is best to visit a company within a similar type of industry and size, but this is not necessarily a requirement. Though this tends to add credibility from the perspective of many, the ERP pitfalls are not necessarily unique for different types of industries or company size. First, look for companies convenient to visit.

Obviously, meet with a company successful with ERP since you want to learn what they did right. At the same time, *what went wrong* tends to get the most management attention. This means the perfect ERP success story is not necessarily best for educational purposes.

In preparing the visit, it is critical to develop a detailed agenda. Start with the formal education topics previously discussed. From this point forward, most education builds upon those major themes.

When developing the agenda, try to involve a broad representation of the host's former project team to obtain different perspectives. When possible these include the project manager, team members (from the functional side of the business), and IT support.

Senior Management and the Software

Though usually not very useful for an executive site visit, at some point it is important for senior managers to be exposed to the new system. Every executive has a different tolerance or aptitude for details, but they need to understand key capabilities of the package at a high-level. What the software can and cannot do might have major implications on daily operations, customers, expected benefits, and enable new business strategies not previously envisioned.

An executive overview presentation or a high-level software demonstration by the recommended vendor is appropriate as part of the final executive sign-off on the package. During the project, targeted demonstrations can play a part in continuing education and sustaining their involvement. The functional analyst on the project team, not the vendor or consultants, performs these demonstrations.

For example, software demonstrations are a great tool to facilitate discussions in many areas such as what must change from a business standpoint and why the change is necessary with the software. This positions the steering team to guide project team decisions and support the changes.

The Role of Firm Practice Leaders

As mentioned, one role of the project manager is to manage executives, but the software consulting firm engaged should assist in this area. Beyond the involvement of a project management consultant, the firm partner or practice manager should be active with the steering team from the beginning. Usually the first challenge for firm management is to take off their sales hats and become project directors.

While it is not a requirement that a firm partner attend every steering team meeting, this person should at least attend the early meetings to help get the project started correctly. Also, involvement at critical junctures (such as the launch of major project phases) helps ensure that the steering team understands the important issues and decisions at each stage. Though partners of most firms can charge very high hourly rates, you should not have to pay for their services (as explained in Chapter 6).

Supplemental Education

Supplemental educational materials and events can add value in reinforcing education, and they are usually available for little expense. For example, nontechnical and practical books about ERP (like this one) should be required reading for the project management team and selected chapters for the executive sponsor and others on the steering team.

Articles, web casts, seminars, on-line readiness assessments, etc., are good resources, but be very selective. Do not over do it, and avoid those that are a sales pitch in disguise. Again, the key word is independent sources of information.

Ready for a Readiness Assessment?

At some point, the question of a *readiness assessment* will surface. A readiness assessment is an independent audit of the organization and/or project that can occur at different stages and have different objectives.

For example, an *ERP Readiness Assessment* focuses on various aspects of the company including current processes, resources, technology, and culture. The goal is to communicate to management the actions necessary to prepare for the project. Therefore, it is a form of education with specific recommendations.

Of course, every ERP readiness assessment consultant believes every project needs a readiness assessment. As an analogy, if you listen to the insurance industry long enough, you might believe every American needs all forms of insurance. The need for an ERP readiness assessment is similar to determining the need for certain types of insurance. You should evaluate the risks versus the cost.

No doubt, there are ERP failures, so it is not a myth that companies must get educated and manage project risks closely. However, some consultants attempt to strike fear in the hearts of their clients in order to sell them expensive readiness assessments and other services they probably do not need.

When it comes to any type of assessment, the great majority are not necessary if you do the other things correctly. The steps discussed previously to educate management can make an ERP readiness assessment somewhat redundant. This approach is normally less expensive and more effective than consultants billing for weeks and then handing senior management an assessment report. Why? If for no other reason, senior managers are involved in the discovery and learning process.

Finally, what are you paying your software consultants to do? Once on board part of their *normal* job is to help the company get ready. Why should any organization pay their existing consultants (or another firm) more money for a separate assessment?

I recommend a readiness assessment only as a last resort to prevent the project from steering off course (and there are usually early warning signs for this). Nevertheless, if it helps a project manager sleep at night and there is plenty of money to spend, rest assured that a readiness assessment has never hurt a project.

CHAPTER 5

SELECTING ERP SOFTWARE

Do Not Doom the Project from the Start

Selecting the right ERP software package is certainly a plus, but by no means a guarantee of success. We see proof of this all the time when two similar companies within the same industry implement the same software with very different outcomes. Picking a software package is easy, but implementing any package is the hard part.

In the best-case scenario, the hope is that the new software will address the majority of needs or, better yet, become a selling point with users because of its great features and functions. Nevertheless, ERP systems with good functionality have been around for a while, and many organizations have made software modifications for the better. Therefore, software functionality as a major success driver is not as powerful as it once was.

When evaluating software, recognize that no package is a perfect fit, and no evaluation process mitigates all risk, no matter how much we analyze it. Perhaps a more realistic expectation is that the new software does not become a liability. The project team has plenty of other things to worry about without making software one of them. Successful projects are at least *software neutral*.

On the other hand, selecting the absolute *wrong* package will be difficult to overcome. This is true even when we do everything else right. If the software becomes a drag on the project for whatever reason, these issues are rarely within ability of the organization to address. This is the main reason it is important to do software selection right.

The good news is there is a low risk of making a wrong decision when the software selected:

- Adequately addresses the high priority software needs.
- Is installed in many companies within your industry segment.
- Is utilized by at least some organizations with more users (total and concurrent).

- Is mature, not full of "bugs" or on the "bleeding edge" of technology.
- Is fully supported by the supplier, and the end of support is not right around the corner.
- Is developed by a vendor whose direction and focus for future functionality are consistent with the needs of the organization.
- Has consulting services and formal training readily available through the vendor or third party resources.
- Is provided by a supplier that is financially viable (not going out of business tomorrow) and not on the verge of being sold to another vendor.
- Is affordable—this includes the ERP software, supporting infrastructure, implementation cost, and ongoing support. Whether many like it or not, cost is always an issue.

Evaluation Areas

Any evaluation should address software functionality (fit), technologies, implementation support, vendor software support, and vendor viability (long-term). The basic software selection steps include the following:

1. Document the project assumptions (objective, scope, budget, etc.).
2. Develop a list of must-have software capabilities and a list of detail requirements in each area.
3. Determine the vendors (packages) for which to request information (RFI).
4. Prepare the RFI and send to vendors.
5. Review the vendor's written responses to the RFI.
6. Determine the short list (two or three packages to evaluate in detail).
7. Receive the first serious quote from vendors on the short list.
8. Perform the detailed evaluation in all areas.
9. Quantify the evaluation results for each package under consideration ("package scoring").
10. Check references.
11. The team makes a software recommendation to senior management.
12. Negotiate the software licensing and support contracts.
13. Obtain senior management approval.

No Shortage of Package Information

Perhaps more than any other ERP topic, information and tools from independent sources are readily available to evaluate software. This includes the lowdown on just about any package—major functionality, technologies, and forward-looking strategic opinions relating to the package and vendor.

In addition, templates are available for RFIs, industry requirements, software demonstration scripts, and package scoring. Take full advantage of these free or relatively inexpensive resources.

Software Selection Consultants

It is also worth noting that the ability to make a sound software choice with fewer (or no) consultants is more doable than ever before. But certainly, get any help required since some consultants are aware of the strengths and weakness of packages for particular industries and software tiers (to the level not always available in the industry rags).

However, history has shown that selecting ERP software is one of those areas where having more consultants does not necessary mean making a better decision. This is because selecting software is not a science, as many software selection consultants want you to believe. The final choice always involves a certain amount of unknowns. No one will know for sure if a package was the right choice until you try to implement it.

In order to do more on your own, look to the important in-house skills previously discussed. This includes project management, business analysis, and IT support. We need a leader and a few analysts on the selection team that can use a somewhat structured evaluation approach, define the business needs, facilitate meetings, and understand the technology choices.

When software selection consultants are necessary, again, make sure they are independent of any vendor. This seems obvious enough, but many hire software selection consultants that provide implementation services for a package under consideration. What software do you think they will recommend?

In addition, like any other type of consultants, focus them on areas where they add the most value. Remember, even less experienced consultants can "bill out" at a rate of over $175 per hour.

Also, many evaluation steps do not require a great amount of expertise and can be accomplished internally with some guidance from a consultant.

Examples include researching packages available, scheduling meetings and software demonstrations, gathering evaluation results, preparing spreadsheets, and all the documentation and coordination an evaluation entails.

When consultants are required on a limited basis, bring them in at critical points for guidance and validation. This requires only a few days of consulting support at each stage versus months associated with the blanket approach. The critical points at which to engage consultants include:

- Review the list of must-have software needs.
- Review the list of vendors to request information.
- Review the list of vendors to evaluate in details (the short list).
- Review the list of detail software requirements and demo scripts.
- Validate the two package finalists' considering your requirements.
- Review the team evaluation results and recommendation.
- Review the contracts and your proposed amendments.

Eyes Wide Open

The purpose of this chapter is to focus on the right approach and the pitfalls that can make or break a software decision—not to walk through the nitty-gritty mechanics of evaluating software.

Every package and vendor has associated risks. When it comes down to it, evaluating software is about identifying these risks. The goal is to understand the strengths and weaknesses of each alternative to make an *informed decision*. As mentioned, it is impossible to identify all risk and unknowns. The idea is to conduct a thorough evaluation and select a package that narrows the window of risk to a tolerable level.

Playing the Shell Game

Anyone that has evaluated ERP systems understands software vendors can get in the way of informed decision-making. Many attempt to turn the evaluation into a beauty contest or a game of hide-and-seek, or they try to get their potential customers to make an emotional decision. If there ever was a time to play skeptic, it is when evaluating ERP software.

The *Show Me* Philosophy

As of last count, there are over one-hundred ERP software vendors out there, and they all have "the best solution." Therefore, you cannot assume much of anything. It is almost to the point where all vendor claims must be challenged for hard evidence.

The philosophy must be, "Unless you can show me, I must assume the capabilities do not exist." This does not make one a popular person with the sales people, but unfortunately, vendors have left us with no other choice. In the end, if the software claims are false, you have more to lose than the vendor does.

Also, even with the best intentions, software vendors do not really understand the specifics of your business. Sure, they ask questions and may send in their team of "functional experts." Many of their experts are really more sales people. Even if they grasp the issues within your company and offer solutions, this does not mean their software addresses any of it.

Get Management's Skin in the Game

When evaluating software, all too often senior management sits on the sidelines awaiting a software recommendation. While we certainly do not want them picking a package, we definitely want their skin in the game.

It is important to realize senior managers are stakeholders, not bystanders. If the software does not fly, they are ultimately responsible. Their involvement also promotes their ownership in the software decision, which is good to have on your side. Moreover, management might have to sell the project to secure corporate funding; it is best if they understand what they are selling.

Senior managers add a strategic perspective to the evaluation in terms of where the business is going, key priorities, and high-level software needs. These typically include capabilities to address business plans, new operating philosophies, key improvement areas, compliance issues, and management reporting. Some of these needs might make the list of must-have software requirements.

At the same time, never let senior management get away with a laundry list of conflicting, pie in the sky software wishes. Glaring contradictions in directions or unrealistic expectations must be resolved now.

Another way for executives to participate is for them to assign a *strategic weight of importance* to each software area or module. As discussed later, the weights can directly affect the package recommendation. Strategic weights recognize that some areas of the business are more important than others are. For example, in a distribution business, chances are Warehouse Management is more strategically important than Accounts Payable. Only senior management can make this determination.

When management defines strategic weights, require a *forced* percent allocation of the available 100% across the modules. This requires them to discuss issues, priorities, and reconcile differences.

Software Selection: A Change Management Tool

While selecting the *best fit* software is the ultimate goal, who is on the evaluation team and making the recommendation can have a major impact on user acceptance. The first mistake is when upper management mandates a certain package. This *top down* approach can foster considerable resistance from those who must use the new system.

Employees want a voice in important decisions that affect their jobs. While running a business is not a democracy, senior managers are usually not in the best position to be selecting software. Their role is to provide input to the evaluation, ensure due diligence on the part of the team, and approve or reject the team's software recommendation. Sending the team back to the drawing board is fine, but software edicts are a different story and add risk to any project.

In addition, the person managing the evaluation process speaks volumes to the rest of the company. A creditable manager from the functional area (business side) should lead the team. The role is to plan and facilitate the process to select a package that best meets the needs of the organization—not just the needs of a given department.

Leadership from the user community creates a greater stake and ownership in the software decision. When there is user buy-in, they will have a higher propensity to overcome obstacles during the implementation. This is the case even if the software later proves to be less than the best choice.

Finally, when the right end-users are on the selection team, this increases the likelihood that their peers will accept the decision. Never underestimate the influence some non-managers have over other employees. When we fail to involve the doers, the word on the street is that management selected the wrong software. Whether this is true or not, it is noise we do not need.

The other mistake is when the IT group or outside consultants are leading the charge. We want to avoid perceptions that the software was pushed upon users by those that do not understand the business or have a different agenda.

Also, when IT runs the selection, the project could be mistakenly viewed as just another technology initiative.

For sure, the IT department must support and assist with the evaluation and is directly responsible for portions of it. These include evaluating technology, assessing vendor support, and perhaps helping users define their needs. But this is different from leading the initiative or determining if the software addresses business needs. Both of these are the job of the users.

The use of consultants to coach, assist, and validate is fine, but the issue is similar to IT leading the selection. It is no secret that many employees view consultants as outsiders (who do not understand the business) and sell crazy ideas to a gullible senior management. In this case, no matter what package is selected, it will be viewed as consultant driven.

Do Not Let Vendors Hijack the Evaluation

ERP software vendors will attempt to exploit every opportunity that provides them an advantage. First, never let vendors define your requirements, the evaluation steps, meeting agendas, or influence any other factors that will distort the evaluation. If this happens, the team will be comparing apples to oranges and a great deal of bias and subjectivity will enter into the equation. This is when software selection becomes an emotional decision.

On the other hand, some bad sales people might have some very good software. Look beyond people, personalities, and presentation. Remember, the goal is to select the best software, not the best sales team. Still, do not ignore obvious red flags.

Staying in control of the evaluation requires a structured approach and sticking with it. Furthermore, the evaluation leader must ensure a level playing field at all times. If there is a decision to deviate from the evaluation steps for one vendor, it should deviate for all vendors.

The evaluation leader must act as a gatekeeper to limit vendor access to senior management, the selection team, or other decision makers. When vendors are permitted to run wild, we might as well hand over the evaluation to them.

The Law of Diminishing Returns

While due diligence is a necessity, many take it to extremes and treat software selection as some mysterious science project. In spite of this, many would have been better off selecting software out of a hat.

While no one suggests prematurely jumping into a package decision, there comes a time in any evaluation when we are simply trading off one set of package issues for another—none of which affects the final decision or changes any package for the better. In addition, many of the issues identified during the evaluation will become non-issues later, and new issues will surface in areas initially believed not to be a problem.

This *law of diminishing returns* is worth noting since many organizations spend over a year and over $400,000 evaluating software, and then want it implemented in six months or less. The extra time and money spent beating software vendors and packages to death could be better used addressing other things that make a big difference. These include educating management, redesigning business processes, managing change, and acquiring more in-house software knowledge.

The majority of the information ultimately driving the final software decision is known after analysis of independent research, vendor responses to the RFI, and the vendors on the short-list demonstrate their software versus your must-have requirements. Of course, take the necessary time to close the risk gap, but any selection process that takes more than six months is taking too long.

The Wrong Software Selection Drivers

Selecting ERP software based on cost alone is the most obvious example of the wrong decision driver. The other is technology. There is plenty of history in this area—much of it not good.

Back in the 1980s, IT department mandates to use their preferred proprietary mainframe technologies drove many ERP software decisions. This was largely about the technologies the IT department could or would support, not the application software that best satisfied business needs.

As a result, some organizations put themselves on a technical island, down the path of early obsolescence or took a step backward in terms of software capabilities. An IT department that does not want to relinquish certain technologies or update their skills should never be the sole reason for purchasing a particular package. However, there is another side to the story, and this one is about not being conservative enough.

Later in the 1990s, the *mainframe* versus *open system* (client/server) wars within the industry caused many to take a blind leap of faith into open systems only to find out later that the software packages in this arena was not as mature as those on the mainframe. Though open systems eventually won, many jumped head first into this brave new world simply too early.

Today, software-as-a-service (SaaS) and cloud computing deployment strategies represent the same fork in the road. No doubt, tomorrow there will be a new one. Nevertheless, do not ignore the lessons of the past: *When the software functionality does not do what we need it to do, nothing else really matters.*

Business needs should drive technology decisions, as long as it does not put you on the bleeding edge. New technology is great, but only if it solves business problems or does not create new ones. Technology can be a double-edged sword, so my advice is to stay within the mainstream of choices available at the time.

Obviously budgets are not unlimited, technology can be a strategic enabler, and there are other important trade-offs. But if the software is a bad fit, everyone will forget all the seemingly valid reasons a package was selected in the first place (cheaper, new technology, IT said so, or everyone else is doing it). Instead, they will focus on the lousy functionality. Software decisions that do not weigh functionality the most can defeat the purpose of a new system.

When all else is equal, technology advantages or disadvantages can tip the scale. Otherwise, as long as the cost of ownership is affordable, the technology stable, and the software is supported, go with the package that best addresses functionality needs.

The SaaS Option

Today, there are two viable software acquisition strategies to choose from—SaaS or purchasing the software licenses outright.

With SaaS, customers own the data but "rent a system" hosted by a provider. Depending on the arrangement, the customer might retain some system administration functions (configuration changes, security, etc.) or turn the process of making these changes over to the vendor.

When purchasing the ERP software and IT infrastructure outright, there are numerous support options. For example, one might host the system internally or outsource it to a third party that maintains your system. As is the case with all outsourced systems, the vendor may support the entire system or certain components only such as the hardware, operating system, and database. In the latter example, the client administers the application.

SaaS versus the traditional deployment methods is about trade-offs, depends on the organization, and is an IT strategy decision that goes beyond the scope of this book. However, my prediction is that ten years from now, any IT department hosting an ERP system internally will become a dinosaur.

The relentless focus on IT cost reduction will continue to drive the adoption of the SaaS alternative. During this time, the functionality of upstart SaaS packages will improve, established packages will continue to move to SaaS, security concerns will dissipated, and the cultural resistance within IT departments comes crashing down.

Today, SaaS solutions are appropriate for many, especially those requiring basic functionality or new to ERP who want to avoid the up-front IT investment. In addition, as traditional Tier 1 and 2 vendors port their applications to SaaS, it becomes much more appealing to larger companies, since these packages tend to have more robust software functionality.

Nevertheless, those late to the SaaS party will be large enterprises or those with complex IT environments. The reasons are legacy system integration with SaaS and a culture of making software modifications.

For example, most large companies have many applications, usually a core ERP system, some bolt-ons, and even some home grown stuff. Most of these systems are currently well integrated or interfaced. Even if they wanted to replace everything with SaaS, these types of enterprises cannot go Big Bang or anything close to it. Hence, temporary interfaces to legacy systems are unavoidable during perhaps a lengthy transition. Also, those who heavily modify their ERP software will delay moving to SaaS since it focuses more on using standard software functionality.

In addition, some of these applications fulfill very specific needs and many companies will want to keep them for a long time; thus, permanent interfaces are necessary. Of course, interfaces are nothing new, but most large or complex businesses have spent many years developing them and do not look forward to doing it all over again (especially interfacing with software in the clouds)!

Though there are new middleware technologies to integrate with SaaS, most IT people know system integration is never as "seamless" or "real-time" as vendors advertise. This is the main reason it will take ten years for SaaS to become the normal ERP deployment strategy.

Planning Assumptions, Risks, and Constraints

Obviously, by this time some thought has gone into the project. However, prior to evaluating software and entertaining vendor quotes, it is time to start documenting project *assumptions, risks, and constraints*.

Assumptions represent the current thought regarding the project scope, resource commitments, the software rollout strategy, the implementation methodology, and amount of consulting support necessary.

A *risk* is any potential event that might adversely affect the timeline, cost, or benefits. This book is all about mitigating risk, but there are probably some unique to the project that should be identified and addressed in the plan.

A *constraint* is any limitation place on the project or team by senior management or any entity external to the company. Constraints *will* affect time, costs, or potential benefits. Senior management and the project management team should identify project constraints.

The list of planning assumptions, risks, and constraints helps ensure all vendors are quoting based on the same understanding of the project. The list is maintained from this point forward since it is the basis for early purchasing decisions and later is input into the detail planning process. Therefore, we want a solid list to start, but working with vendors during the quote process helps fine-tune the list.

Two-Levels of Software Requirements

Software requirements are a fundamental tool to evaluate and select software. The organization should develop the requirements. When help is necessary from a consultant or when using industry templates for example, make sure they are your requirements (not someone else's needs).

Most evaluations have two levels of requirements. The first is a list of *must-have* software needs. Use these to determine the vendors to include in the request for information (RFI), to evaluate vendor responses, and to determine the packages to evaluate further (short-list).

The second level is a list of *detail requirements* describing the software functionality needed within each application area (module). These requirements are many, and are used to communicate specific needs to vendors on the short-list and to evaluate the functionality available within each package.

The Must-Haves

The must-haves are a high-level list of about twenty items representing the most important needs of the business. If a package cannot address the great majority of the must-haves, there is no point in evaluating it further.

These requirements can come from several sources. First, focus on the project objectives and expected benefits, and the type of software capabilities necessary to achieve them.

Second, there could be important functionality in the current system that must exist in the new system. Third, as discussed previously, senior management might add a strategic perspective in formulating important business needs. Moreover, when analyzing the detail requirements, some will resonate to the must-have list.

Finally, similar types of companies and industries have somewhat common requirements, at least at the high-level. For the purposes of the RFI and determining the short-list, this information is useful.

For example, the ERP industry in general has classified packages based on support for different *industries, operation strategies, organizational structures*, and *types of customers* served. Review the classifications for various packages and create a list of the major capabilities or attributes applicable to your organization.

Industries (also referred to as *verticals*) might include manufacturing, distribution, construction, public sector, property management, etc. Using manufacturing as an example, industries can be further categorized as fabricators, food and beverage, pharmaceutical, consumer packaged goods, aerospace/defense, etc.

A different angle is the operational strategies utilized within a company. In the manufacturing example, there are companies that engineer products to order, assemble products to order, build to stock/forecast, build to order, or a combination of the above. On the shop floor, there is discrete, batch, repetitive or mixed mode manufacturing processes. The key is operational strategies are not necessarily unique to a particular manufacturing industry.

Another attribute to consider for the list of the must-haves is the structure of the organization. Often, this is referred to as the "multiples." For example, an enterprise might consist of multiple companies (from a financial standpoint), multiple plants and warehouses, and have the need for software that supports multiple languages and currencies.

Finally, the type of customers the business serves also shapes the must-have list. Companies selling directly to consumers have some different needs than those selling to retailers or the government.

For example, when doing business with the federal government contract traceability is a very important need. In addition, different methods to interact with the customer are necessary for those that sell directly to consumers versus to retailers. Perhaps mobile or web based front-end applications for consumers and EDI for retail customers.

Flush out the Detail

The list of detail requirements represents the specific business needs and/or software capabilities required. Consultants may or may not be necessary to assist with this task, but the business analysts and users on the team should take the lead in gathering, analyzing, and documenting these requirements.

Since ERP packages are designed to support business processes, it makes sense to organize these requirements around the processes associated with each software module.

When developing requirements, do not attempt to list every conceivable line item the software must support. For example, every ERP package should allow for printing of customer invoices or purchase orders. Instead, focus on the unique requirements in all areas and the important ones that drive the business.

The list of requirements is never all-inclusive, because there are many unknowns. The good news is it does not have to be. At this stage, we are trying to select the best software tool, not determine specifically how it will be used.

There are several methods to identify requirements for evaluating software. When combined, they provide input to the business analysis necessary to develop a clear list of needs to present to short-list vendors.

1. **High-Level Business Analysis**
   An analysis of current processes and needs is one method to assist with gathering requirements. The best time to start this analysis and how much detail is necessary are always topics of debate within the consulting industry. However, the sooner the analysis begins, the

more that will be known before making important project decisions or a financial commitment of any kind. The very best time to start is before evaluating ERP software.

In terms of the level of detail, at some point, mapping the current processes (the *as-is* analysis) is recommended. However, for the purposes of selecting software, it is usually not necessary to construct detail business process maps. Typically, analyzing the major workflows and brainstorming needs with a cross-functional team is enough for selecting a package.

2. **Interviews or Surveys**
    These shed light on the important needs specific to each department and include interviews or surveys with functional managers and key users.

3. **Review of Legacy Software Functionality and Limitations**
    This is a coordinated walk-through of current system transactions. The purpose is to identify capabilities that are important to retain or to identify potential improvements from the end-user perspective. Also, the IT department can usually provide valuable insight during this analysis.

    Of course, just because certain functionality exists in the current system does not imply that it is required in the new system.

4. **Software Requirement Templates**
    Available from independent sources, these describe the detail requirements typical of various types of organizations and industries. Templates should be used only as a starting point or to help round out the requirement list.

5. **Best Practices Requirements**
    When planning to implement an industry *best practice* the problem is how to know what is required when we have no previous experience with it. This is where project team education comes into play. This is important since there are probably benefits in the

project justification that assume the best practice will materialize (and the software supports it).

User "Desirements" Versus Business Requirements

There are several traps to avoid when defining software needs. First, simply asking a user what they want may not identify what is truly required. We do not want to develop a vague wish list of individual user "desirements" shaped solely by the current environment. This is another area where business analysis skills are important.

For example, many managers view information technology as a convenience tool and want to automate just about everything. Usually they have good intentions, but this mentality ignores the fact that some processes are better performed manually, some decisions are best made by people, and no package automates all activities.

In addition, many users live within the context of the here and now. If you ask them what they want, the answer is "exactly what I have today." Sometimes this is about employees not wanting to change or the desire to make their transition to new software as painless as possible.

Workaround Requirements

Perhaps a bigger challenge is when software is viewed as the easier alternative to addressing the real business issues. This is when the dysfunctional aspects of the organization shape the software requirements.

Automating creative ways to *workaround* the root causes of process deficiencies never helped any organization. Workaround activities always take more time to perform and cost more to develop. This is why we need a cross-functional approach to gathering requirements, in order to get at the source of the process problems.

When extensive software modifications to the current systems are the norm, this is when automating workarounds can become part of the company culture. ERP is usually more difficult to implement in these types of environments because it is harder to contain requests to modify the software. In this case, you must select a very robust package and will require a very strong statement from senior management that discourages requests to modify the software.

Software requirements that are based on workarounds leads to one of two outcomes: Discovering that no package meets your requirements (and for good reasons) or experiencing costs and schedule overruns to make unnecessary software modifications.

Define the *What,* Not the *How*

When purchasing any system, one must acknowledge that specific software transactions and procedures will change. There is no way around it—the look and feel of the new system will not be exactly like the old system (even if everyone loves the old system).

When writing software requirements, focus on *what* the software must do, not *how* it should do it. Defining the *hows* limits the flexibility of any package to address the requirements since usually there is more than one way to accomplish a task. Searching for the *hows* can again lead us to the false conclusion that no package is a good fit, but this time it is because the requirements are too rigid.

How a package accomplishes a given function is properly reflected in the "score" assigned during the evaluation. The score represents the overall software fit to the requirement. Scoring is typically zero through four, with four being the best possible fit. For example, if a package satisfies a requirement regarding what it must accomplish, but there is a concern about how it accomplishes it, the score assigned should be less than four.

Weight of Importance

As discussed, senior management should assign a strategic weight of importance to each software module. This reflects the module's relative importance to the overall success of the organization compared to other modules.

In addition, weights apply to *individual requirements* within each software module. Each evaluation team should reach consensus on the weight of importance of each software need within their assigned module.

Later, the weight of importance is used as part of the calculation to rate each package in terms of *software fit*. For example, the weight of each requirement within a module equals:

3 = High Importance

2 = Medium Importance

1 = Low Importance

Request for Information

The RFI enables a consistent format for vendors to provide information and to evaluate their responses. The overall goal of the RFI is to determine the two or three packages to proceed to the next phase of the evaluation (the short list).

When at least two packages successfully address the must-have requirements and most other RFI items, the risks associated with the final choice (after the detail evaluation) are greatly reduced.

The RFI is intended to cover all the important evaluation areas without getting lost in the details at this point. The goal is to achieve a broad perspective of each package and vendor to determine which candidates are worth further investigation.

By focusing on the bigger picture first, *showstoppers* associated with any package are identified earlier. Therefore, a good RFI not only reduces the risk associated with the final choice, but also helps expedite the selection process.

When preparing the RFI, there is no need to write a novel or something that reads like a marketing brochure about your company or project. Avoid the fluff and get down to business. Some of the best RFIs are on a simple spreadsheet, three or four pages long.

The RFI outline below communicates to the vendor basic information about the company and project, additionally providing a place for the vendor to record responses to important questions.

Other than the must-have requirements, there are very few RFI items unique to a specific organization. The questions represent what any company should be asking. The only exception is when considering software-as-a-service. In this case, there are additional considerations mostly addressing technologies and classic system support functions otherwise provided by the internal IT department.

I. Brief Description of the Company

    a. Products, services, key customers, number of employees, physical locations, current systems, and number of system users.

II. High Level Project Descriptions

    a. Business case, objectives, scope (modules, key business processes, etc.)

III. Description of Must-Have Requirements (No More than 15-20)
This was previously discussed.

What to look for: Vendors should describe how their software addresses each requirement. A response such as "no" is fine, but a simple "yes" is not a good answer. We need some idea of what the system does to support the requirement.

IV. Basic Package Information
- Date of initial software release
- Number of major releases since initial release
- Total number of customer installations
- Total number of installations within your specific industry type
- Notable customers within the industry type
- Number of total and concurrent users for largest install
- Access to source code (yes or no)
- Deployment options (SaaS or internally hosted)

What to look for: This is an excellent place to quickly weed out packages from consideration. All the items listed here are indications of the overall state of a package and general applicability. For example, if there are very few installs of any type, this is a concern. In addition, if your company *builds-to-stock* based on a sales forecast, but the software is installed mainly in *engineer-to-order* companies, what does this say about the software fit? Also, if your organization will have 300 concurrent system users, but the largest customer install is 150 concurrent users, what does this say about the potential for system performance issues? No access to the source code means software programs cannot be modified by the customer. The vendor might provide customer specific modifications or they might not. Even if they do, vendor provided modifications are very expensive. Finally, the date of the initial software release and number of subsequent releases are an indication of package maturity. If the software has not been around for a while, it probably falls short in terms of functionality and likely has more than the average number of bugs.

V. Technology

- Application architectures (web-based, mobile, thin client, etc.)
- Operating systems supported
- Databases supported
- Interfaces supported
- Major technology changes within current release or planned with next release
- Software Development (programming) Tools (note as proprietary or third-party)
- Servers, data collection and other hardware supported (note as proprietary or third-party)
- End-user data analysis and report writers (note as proprietary or third-party)
- Third-party packages purchased and incorporated into the base offerings (e.g., core software modules not developed by vendor)

What to look for: Any company requires technologies that are reliable, fully supported (by both the technology vendor and the ERP provider), and cost effective to purchase, support, and grow (scalable). With exception to software development tools, any vendor proprietary software or hardware is not the best-case scenario in terms of cost and integration flexibility. In addition, any package that is really a bundle of independently developed software (acquired by the vendor) and thrown together as a single ERP offering is of concern. Finally, when hosting the system internally, the more technology options associated with any package, the better.

VI. Implementation Support

- Number of project managers and application consultants employed by the vendor
- Percentage of project managers and application consultants that are subcontractors (not permanent employees)
- Number of vendor project managers and applications consultants located within the region

- Name of third-party firms providing implementation/consulting services
- Name of third-party firms providing formal software training services
- Vendor training delivery methods (web-based, vendor site, customer site)

What to look for: More project management, application consulting, technical, and training options available from any source (vendor or third party) means more flexibility. The number of consultants employed directly by the vendor is an indication of their level of commitment to the implementation side. In other words, is the vendor primarily in the business of just developing or selling software? Local versus non-local consulting support is not a deal breaker at this stage, except perhaps if the consultants are located halfway around the world.

VII. Vendor Software Support

- Description of each type and level of support provided (maintenance program options)
- Number of customer service support personnel
- Number of technical support personnel
- Number of internal software developers
- Number of sub-contracted software developers (not permanent employees)
- Number of software developers assigned to sustaining support/fixes
- Number of software developers assigned to new releases
- Overview of enhancements in the development pipeline
- Vendor provided software modifications for customers (yes or no)
- Support for software modifications performed by customers (yes or no)
- Release date of the current version of the software
- Scheduled date of the next major release
- End date for new development (new functionality, new bug fixes, and technology upgrades)

- The frequency in which customers are required to upgrade to new releases
- End date for all support
- New ERP package in development to replace the package (yes or no)
- Other products, services, or packages supported

VIII. SaaS Considerations
- System security strategies and tools
- Data backup strategies and tools
- Interfaces provided to customer legacy systems
- Disaster recovery contingency plans
- Software configuration changes by the customer (yes or no)
- System performance and availability

What to look for: The availability of several different software maintenance programs enables companies to better match the level of support with business needs. The number of resources allocated to various areas of support is an indication of vendor priorities. The frequency of new releases tells much about the availability of new functionality and the vendor's commitment to the product. The type of enhancements currently in development is an indication of the overall direction of the software. For example, is the planned functionality something you can use or is the vendor positioning the product to support an entirely different industry? Finally, every product has an end-of-life. While most ERP vendors have a *product roadmap* with end-of-support dates for specific versions of the package, most do not publish end-of-product *life* dates. But considering other packages or products in their development pipeline, this is an indication of where the package is heading. It could be headed for the scrap heap or vendor support becomes diluted in order to shift focus to other product priorities.

IX. Vendor Viability
- Year of incorporation
- Public or privately held

- Financial information request
- Dun & Bradstreet

What to look for: In the world of technology, the landscape is constantly changing. The viability and staying power of any vendor is always of concern. The idea is to select a vendor that is going to be around for a while. The rise of SaaS makes this area more risky than in the past since it is harder to pick the winners and losers. In this space, some new vendors will not make it, but others will exploit the opportunity and leap the competition. Some of the previous winners in the old world order could become losers in the transition. In the meantime, a vendor that is financially unstable or ripe for a takeover is seldom good news for customers. Even if another vendor acquires the package, this can be very disruptive during an implementation. In addition, no one knows for sure if the new vendor will continue enhancing the package or just milk customers for the maintenance fees. In this area, listen to the word on the street, including the industry rags, ERP think tanks, and even competing vendors. Though most vendors paint a grim picture of their competition, they can have a few pieces of worthwhile insight.

X.   **Pricing (ERP software and consulting rates)**
- Method for software licensing (named users, concurrent users, servers or module based, etc.)
- Initial software price quote (given project scope assumptions)
- Annual software maintenance fees as a percentage of software cost (for each maintenance program option)
- Application consulting price per hour
- Project management consulting price per hour
- Technical consulting price per hour

What to look for: Though software functionality is king, the cost to purchase and support a package does matter. For the purposes of the RFI, rough quotes for the software, consulting hourly rates, and maintenance fees are a good indication of both the short and long-term affordability of the system. The price for annual maintenance can be as high as 25% of the software cost, but like the software, is usually negotiable.

The Vendors to RFI

Considering the sheer number of ERP systems available in the market, it does not take long to realize the list must be reduced to a manageable level. When determining the vendors to RFI, the goal is to find good reasons not to consider them. Look for major showstoppers in all evaluation areas to identify the five or six vendors to RFI.

The first major considerations include: The must-have requirements, cost constraints, the deployment options (SaaS or traditional), and where your company falls within the so-called *software tiers* (discussed next).

There are many independent websites providing this type of information. These can be used to start the process of elimination. No doubt, one will need to call a few vendors to get more information, but identifying the five or six vendors to RFI should not be difficult or take forever.

Software Tier Implications

When exploring ERP options, you will find that vendors and industry research firms developed the concept of *Tiers* to categorized ERP vendors and packages. This information can be useful when determining the vendors to RFI.

From my perspective, each Tier implies a different ERP market based on company revenue, total number of users, software costs, and the scope and depth of functionality. Tier 1 includes the higher-end packages. However, there are several important caveats:

- **There is no direct relationship between a company size and the complexity of its software requirements.** For example, many assume small organizations (SMB) have different, less rigid, and less complex requirements than larger ones. But small does not always mean simpler. There are many examples of small to medium size businesses with very robust software needs. In fact, nothing short of the advanced modules in high-end packages will do. System requirements should be driven by the nature of the organization, business processes, and level of complexity, not necessarily by company size.

In addition, how management views the role of information technology within the business does affect system requirements. For example, every organization has information needs, but that does not mean all needs must be fulfilled by a software package. Management may believe that certain needs can be satisfied just as well outside the system. However, do not underestimate the inefficiencies and inaccuracies associated with off-line spreadsheets and manual activities. When you add it all up, it may be faster and easier to perform the tasks within the ERP system.

- **Degree of business growth or change anticipated.** Where does your company plan to be five years from now? As one moves up the Tier scale, generally the more flexible the software to support various business requirements and growth.

- **The lower the Tier, the harder it can get.** Below Tier 2, the number of available packages begins to explode. Some of these are newcomers with a limited track record. Some may lack the functionality or even modules traditionally found within a full ERP solution. Selecting software is about managing risk, and some lower Tier vendors may either lack financial staying power or offer less stable packages, from a software bug and performance standpoint.

The Short List

First, when a vendor is slow or appears uninterested in responding to the RFI, it could mean their software is not a good fit. Remember, vendors evaluate RFIs and do not want to waste time pursuing business they have little chance of winning. When this appears to be the case, discuss this with the vendor, but do not waste your time either.

The second consideration is, once again, the must-have requirements. Like other areas of the RFI, this requires follow-up discussions with the vendor for clarification of their written responses. At the RFI level of the evaluation, there is usually no need for a software demonstration. If the vendor moves to the short list, this is when the demos begin. But if a package cannot satisfy the great majority of the must-haves, it should not make the short list.

Next, this showstopper mentality should also extend into the non-software functionality areas of the RFI. Major concerns with cost, technology, support, and vendor financial viability should be a cause for early elimination. The last thing you want are users excited about a package during the demonstrations only to tell them later that the package costs too much or the vendor is not financially stable. Get the red herrings off the table during the RFI phase.

The Detail Evaluation

When down to two or three packages (the short list), it is time to expand the scope and depth of the evaluation and the involvement of the remaining vendors. At this point, proper planning and control of the evaluation is critical. The biggest challenge is keeping vendors in check, ensuring a consistent evaluation process for all packages, and minimizing bias and subjectivity within the team.

In addition, the more we can quantify the evaluation the more believable the software recommendation. If nothing else, the numbers provide a basis for the final decision when the choice is not an easy one. The need to quantify the evaluation results is important in the software functionality areas.

In addition to the detail requirements, *demo scripts* play an important role. These represent the day-in-the-life in terms of the basic steps and transactions performed to complete a given business process. The scripts are not the requirements, but are a tool to control software demonstrations and aid in determining if the system satisfies the requirements.

The scripts should be developed by the evaluation team and used by each vendor when preparing for the demos and as a roadmap when conducting them. Otherwise, during the demonstration the vendor will bounce around the system, it will be hit-and-miss, or the vendor will focus on what they want to discuss. In either case, there is more talking and less demonstrating.

Package Scoring

In addition to the detail requirements and the weight of importance, the "package scoring" is the final piece of information to quantify the evaluation results. The score represents the team's opinion on the degree to which a package addresses a requirement *right out of the box*. For example:

0 = Software does not satisfy the requirement.

1 = Software currently does not satisfy the requirement, but the capability is verified to exist in the next release.

2 = Software partially satisfies the need, but workarounds are unclear.

3 = Software partially satisfies the need, with acceptable workarounds.

4 = Software satisfies the need, with no workarounds.

Before getting started, educate the team on proper assignment of scores. Here are some guidelines to use:

- A software modification or enhancement (custom programming necessary) receives no credit (zero score).
- For a *"software change"*, get clarification if this implies a modification or software configuration change. If a setup change, it is best to let the vendor make the changes and demonstrated the functionality again later.
- The need to write a report or if the vendor proposes a somewhat creative procedure to satisfy the need, this implies a workaround. A workaround can either be acceptable, unclear in terms of how it would work, or clearly unacceptable. If unacceptable, it scores a zero.
- *Vaporware* does not count. When the vendor states that required functionality is in a future release but has no proof, give no credit since the software may never support it. But when the vendor can provide ample documentation that the functionality is in the next release, or better yet, you can see it in software under development, give partial credit (1). Never give full credit since the capability is not immediately available and an upgrade is necessary to acquire it.

Thirteen Tips to Manage Software Demos

Other important aspects of managing software demonstrations include:

1. **Do the vendor introductory meeting and basic software navigation first.** This initial meeting is somewhat unstructured. It allows the vendor to get their standard sales pitch out of the way and demonstrate the navigational aspects of the software to the entire evaluation team (instead of covering these topics in each module demonstration).

2. **For each package and module within scope, plan two rounds of software functionality demonstrations and one short wrap-up session.** This should be sufficient to complete the functionality evaluation and tie up loose ends.

3. **Focus first on the important scripts and requirements that drive the evaluation in each module.** These include unique requirements and those weighted highest in priority.

4. **Create an equal forum.** For a given round of demonstrations, the place and delivery method (on-site or remote web-based) should be the same for each vendor. Again, this is all about leveling the playing field. For example, if one vendor performs a demo on-site and another does the same module remotely, who do you think has the opportunity to leave the best impression? This impression may have little to do with the software. Encourage all vendors to be on-site for the demos. When down to the short list, nothing can replace face-to-face communication.

5. **Establish a clear agenda for each session (including a timetable for each topic).** Work with the vendors in establishing the agenda, but do not let them determine it. The agenda should be identical for each vendor for a particular module, at least in the first round. One may deviate in subsequent rounds as specific questions or issues require further investigation.

6. **Request each vendor send their *A-Team* (people that really understand the software and how to demonstrate it).** If a vendor sends their B-Team, this is not necessarily their problem since it can become yours. It is in your best interest to understand what each package can really do. If the demonstrator is not very familiar with the software, this can make a good package appear inadequate.

7. **Make sure all vendors perform demonstrations with a copy of the production version of the software and with technologies**

typically deployed. We want to avoid demos using software or databases not representative of the actual product. Though not common, some vendors modify program code prior to demonstrations to make it appear the software addresses a particular need (i.e., the real package does not do it).

8. **Require the vendor to demonstrate the software release to be implemented (usually the current release).** As a note, upgrading software during the project can be disruptive, and being one of the first companies on a new release adds risk. However, when a vendor wants to demonstrate a previous release this could mean the current release in production is not ready for primetime. Check references of those using the current release, and if what you hear is troubling, run away from the package as quickly as possible.

9. **When one vendor uses sample data representative of your business for the demos, request all vendors to do the same.** Otherwise, this will give one vendor an unfair advantage. Vendors understand the use of your company's data makes their software look less foreign to the evaluation team.

10. **Educate the team on its role and the need to maintain objectivity.** The team's role is to seek clarification from the vendor when responses are unclear. Also, lack of objectivity can get out of hand if not managed. For example, I once sat through a demo where the evaluation team gave the demonstrator a standing ovation when it was over. No, I am not joking. I must admit he was a dynamic entertainer, but the software was not that good.

11. **The evaluation project manager must be an interpreter.** Though the vendor and the team are speaking to each other during the demo, this does not mean they are communicating. Vendors and users come from different worlds, use different terminologies, and the words can get lost in translation. It is the job of the project manager to ensure that communication is truly happening.

12. **The evaluation project manager must referee.** This includes ensuring adherence to the agenda, participation of evaluation team members, and that the vendor is answering the questions and demonstrating the software.

13. **Conduct a team follow-up meeting immediately after each demonstration.** Allow time at the end of each demonstration for the team to discuss what they learned, reconcile differences, score the package, and document follow-up questions. Of course, this segment of the meeting does not include the vendor. When not done immediately following each demo, the team will be hard-pressed to remember what they saw. In addition, two people attending the same demo can walk away with very different perceptions. Independently, each team member records the score for each requirement during the demo. Afterwards, individual scores are discussed and reconciled into a single team score for each requirement. When a consensus score is not possible, it becomes a follow-up item.

Test Drive the Software

Most vendors offer *trial versions* of their software to allow potential customers to get a feel for the functionality. Take advantage of this, but there is a difference between a spin around the block versus attempting a full-blown *proof-of-concept*.

A proof-of-concept involves loading sample company data into the system, making software setup changes, and exploring the software in more detail. This is fine, but to truly prove-out a package, you almost have to implement it.

The other problem is most vendors do not provide free training and support in a pre-sales mode. If they do, the resources are probably not real trainers or consultants and, therefore, may not do the software justice. This can lead to a misguided understanding of the system and unnecessary frustration trying to make it work.

This is not to say a proof-of-concept is never appropriate, but it is more appropriate for packages involving more risk or to verify support for a very important business need.

If the plan is to perform a proof-of-concept, use it as a final verification step for the preferred package. Limit the scope and objectives and insist the vendor provides knowledgeable non-billable resources for at least a short period.

What Is Behind the Curtain?

Users experience screens and reports when working with any system. What is behind the scene usually does not matter to them unless it creates additional work or inconveniences. When evaluating the functionality of any package, do not just take the black box approach. It is important to understand what is behind the curtain. This may uncover problems not easily detected during a software demonstration.

While the hope is that all vendors develop new software capabilities, this functionality is not always as seamless as they claim. In many cases, major pieces of the system were really an afterthought, and it shows.

Anytime a vendor adds new modules or major features into a complex product like ERP, the original design must be retrofitted to some degree. This often entails working around existing programs and data structures in an attempt to incorporate the software changes. This can lead to compromises that negatively affect software capabilities, usability, integration, and real-time information.

If the data to support major functionality was not anticipated in the initial system design, it can be very difficult to add later. Examples include support for multi-businesses, contract management, batch control, or any other functionality or data that touches just about every module in the entire system.

The result of design compromises is that users find themselves entering the same information into two or three separate areas of the package, or behind-the-scenes batch programs are required to keep data in-sync (but usually not real-time). Not a good design.

Therefore, ask the vendor to explain how key functionality is supported within the database. Along with this, review the package's logical data model—every package should have one. A data model is a representation of the *business data* used to support application processing. In business terms, it describes the database, key fields, and how each file relates to other files. This could expose major issues with the underlying system design.

Beware of Bolt-Ons

As an alternative to developing additional functionality within the base package, many ERP vendors purchase third-party software (from another vendor), and sell it as an optional or necessary bolt-on to support a module need.

Though users tend to view the availability of optional bolt-on applications in a favorable light, it can be another example of "software as an afterthought." For example, all vendors say their bolt-ons are fully integrated with the rest of the ERP system. However, get a clarification of what "integration" means because most are interfaced instead.

Integration implies that application programs utilize a single database or information is updated in real-time. On the other hand, an interface is defined as programs that physically extract, transfer, and load data between *separate* databases on a *periodic* basis. Interfaces are necessary since the systems were not designed to work together in the first place.

Beyond separate databases, many times bolt-ons are built on a different technology platform than the ERP software. This makes interfaces even more difficult to design, develop, test, and support. This is one reason interfaces are never fail-proof. When they fail, it can affect the users and business in a significant way.

Going a step further, many vendors offer bolt-ons to their bolt-ons. This is the ultimate afterthought in the world of system design. In fact, any ERP package constructed mainly of third party bolt-ons probably has data redundancies and integration issues everywhere. Do not expect the vendor to disclose this information unless specifically asked.

Furthermore, when a vendor routinely uses optional bolt-ons to provide "point solutions," it could be a sign that the package is in a state of decline within its product life cycle. When this occurs, the vendor is reacting to market needs by purchasing a quick solution versus investing internal resources in a dying product.

When considering a bolt-on, the key is to understand the degree of real-time information required by the business in the application area in question. For example, bolt-ons can be a good solution when providing a specialized functionality, requiring only a loose handshake with the rest of the ERP package. In this case, the lack of real-time integration is not a hindrance to usability.

Evaluating Technology

Generally, I find it unnecessary to quantify areas of the evaluation not associated with software functionality. Typically, a list of advantages and disadvantages in these areas are enough to reach valid conclusions.

As discussed previously, a package should not make the short list if there are major concerns regarding cost, technology, implementation support, vendor support, or the long-term viability of the vendor. At the same time, this does not mean the evaluation in non-functionality areas is over or has no bearing on the final decision.

For example, just because a technology is considered mainstream, does not imply it works well with a particular package. A fairly common situation is getting caught in a *technology transition* with an otherwise good package. This occurs when the software has a reputation for being stable, but the release you plan to implement includes a major shift in technology or the deployment strategy. In this case, past system performance may not be a good predictor of future performance.

As mentioned before, while technology preferences should never be the sole driver of a package decision, it could sway it when all else is relatively equal. For example, technologies that integrate well into the current IT environment (in terms of existing skills and systems) or enable certain strategic advantages from a business standpoint are always helpful (as long as the software functionality is a good fit).

System performance should always be a consideration when evaluating any technology. If concerns about system performance arise, ask the vendor for *benchmark performance* or *volume test* results associated with the package. If they do not have any, this is not good.

Also, cover system performance issues in the reference checks. This is also the time to find out if any particular programs in the system are problematic.

Evaluating Implementation Support

The evaluation of consulting and training services can influence the ERP software decision. At this point, we are assessing the options available, the cost, and quality of each, not selecting these resources.

Review the resumes of project management and application consultants from all legitimate firms associated with packages on the short-list. Request that each vendor provide information only for consultants likely to be available to work on the project. Additionally, at a high-level, assess each vendor's formal training courses and implementation methodologies and tools.

Evaluating Software Support

Now is the time to dive deeper into the types and quality of support for each package in the areas of application and technical expertise, and bug fixes provided by the vendor.

First, it is important to understand the vendor's help desk processes in terms of issue resolution. It is not unreasonable to ask for call statistics for various types of customer issues and resolution times. It is always best if the vendor can provide a documented *issue escalation* process. Otherwise, if there is a major problem it may be difficult to get to the right experts or their management's attention.

Finally, inspect the quantity and quality of software documentation available to customers on their website. This documentation comes in three flavors such as user manuals, white papers (how to configure the application), and technical information.

Percent Fit

There are numerous methods to quantify the extent a package's software functionality meets business requirements. A *percent fit* calculation is easy to use and makes the overall outcome of the evaluation easy to comprehend. For example, "The ABC package meets 90% of our requirements, while package XYZ meets only 70%."

The key inputs to the calculation are item discussed previously including management's *strategic weight importance of each module* (pg. 47), the team's *weight of importance of each requirement* (pg. 58) and the evaluation *score assigned for each requirement* by the team (pg. 67). The calculation below yields a *percent fit for each module* and then the entire package.

MODULE PERCENT FIT

This calculation applies to only the requirements *within* a given software module and equals:

{*The Sum of all Weighted Scores* (weight of each requirement multiplied by the score assigned)} divided by {*The Sum of the best possible Weighted Scored* for the module (weight of each requirement multiplied by the best possible score of 4)}.

Where: The weight for each requirement = 3 (high priority), 2 (average priority), or 1 (low priority). *Where:* The score for each requirement = 4 (highest), 3, 2, 1, or 0 (lowest)

PACKAGE PERCENT FIT

This calculation takes the *Module Percent Fit* (above) and multiples this value by the *strategic weight* (percent) that senior management assigned to each module. Next, these values are summed to arrive at a percent fit for the package as a whole.

Any package worth purchasing should adequately cover the must-have requirements in the RFI and have a package percent fit of at least 80%. Do not get overly concerned if the best-fit package barely meets the 80% threshold. Many software requirements are based on current business processes. It is inevitable that processes will change and many needs will be satisfied in ways not envisioned previously. When no package meets the minimum criteria, it is usually for the following reasons:

- The team did a poor job on the RFI and selected weak packages for the short list.
- The software requirements are tainted to the extent no package will satisfy them (see defining detail requirements).
- The capabilities required within your industry are not well represented in many ERP packages. In this case, expect to make software modifications

Also, perform a *"what if"* analysis by eliminating all weighting and see which package comes out on top. This can get interesting if the non-weighted score suggests a different package. This highlights the importance of getting consensus on the weight of importance of each module and each requirement within a module.

Avoiding Shelfware

When entertaining vendor quotes, plan to purchase only the modules within the project scope. Many vendors like to sweeten the pot by discounting additional software you *might* use later. You not only pay for the software, but also the annual maintenance fees for several years before using it (if ever used, e.g., shelfware). Moreover, if one does not address in the contract the ability to return unused software (discussed later), you might be stuck with it.

Similarly, the way vendors bundle software in a quote does not always reveal the specific components. Require the vendor to break down the quote into the lowest level of licensed components. This may uncover potential shelfware items.

Methods that vendors use to license software varies, but many do it on a *per user* basis. *Users* could mean *total named users* or number of users logged in the system at any given time (*concurrent users*). Also, some have a base price plus the number of users, while others license considering the number of servers. Finally, each component of the software could be licensed differently. The key is to understand the specifics of licensing, because it can get so confusing, some buy more licenses than necessary.

When the licensing method is the number of concurrent users, it is easy to over-estimate needs. As a general guideline, a peak level of concurrent users for most organizations is roughly 40%-50% of the total number of users (named users). The peak usually comes at specific times within the business cycle (usually month-end or year-end). The concurrency levels observed in the current systems are a reasonable guide, but also consider differences in the new software footprint.

Finally, do not inflate the number of licenses based on anticipated future growth. Estimate the licenses required to get the system up and running with some headroom. Treat future growth or business decisions that necessitate more licenses as a separate event, with its own set of costs and benefits. Later, we will discuss negotiating discounts for potential future purchases.

The Cost of Ownership

When selecting a package, we must at least understand the major costs differences associated with the two finalists. Beyond the software licensing cost, the key drivers tend to be technology software, consulting rates, annual maintenance fees, and maybe software modifications. As a point of reference, or to estimate the total cost of ownership, Chapter 13 contains potential cost items.

Normally, the cost of potential modifications are not included when comparing packages unless there are glaring software limitations, management policy does not prohibit modifications, and you would definitely do the mods if the package is selected. In addition, certain software limitations may affect previous assumptions regarding project benefits and the ROI.

This does not imply we should rush off and perform the modifications anytime soon. As the project progresses and the facts become known, the modifications may not be required or may not be approved by management.

Unless one package appears much more difficult to implement than the other, consulting cost per hour is the best indicator of the difference in the total cost of implementation services.

Beyond application functionality, consider what else is included as part of each package. For example, some packages have IT tools and other useful software automatically bundled-in that might otherwise have to be purchased separately from a third-party provider.

Finally, do not assume SaaS is always the lower cost option compared to internally hosted systems. For companies that run their ERP systems for eight years or more and replace infrastructure hardware infrequently, SaaS may cost much more over the long run.

Another cost versus benefit consideration is when moving away from custom programming (more commonly associated with internally hosted systems) to any type of strictly standardized functionality (such as SaaS). Whether we realize it or not, there is an opportunity cost associated with not having the ability to write unique software solutions when truly needed. Alternatively, if the SaaS provider writes custom modifications for their customers, this will not be cheap.

Reference Checks

The final act of due diligence is checking with companies using the top two packages on the short-list (or perhaps just for the preferred package at this point). If the evaluation team has done its job, it is unlikely that reference checks will yield major surprises.

At the same time, a reference check is the last attempt to identify any glaring issues missed previously. If nothing else, it adds creditability to the evaluation. When asked if you checked references, you do not want to say, "No, I did not." This is the first question anyone will ask.

By now, the vendors should have provided a list of other customers within your industry using their software and the total number of system users for each. Of course, vendors push their *marquee* accounts for reference checks, but these should be avoided.

Remember, the objective of a reference check is to uncover issues, and most marquee accounts will not fully disclose them. Most vendors compensate these accounts in some form in exchange for being a reference. Second, many entertain visitors so often that they view themselves as celebrities and do not want to tarnish that image.

A site visit for a software reference certainly does not hurt, but often it is not necessary or worth the cost. Conference calls or web meetings can be just as effective as long as you know the reference actually uses the software!

The recommendation is to speak with three customers in your industry using the same software modules, release number, and technologies planned. Similar to executive site visits for educational purposes, it is best to speak with those directly involved with the implementation, such as the project manager, a functional analyst, and an IT manager. The software vendor should not participate in these meetings since they have no reason to be involved. Send your list of questions to the reference company before the meeting.

The Final Selection and Senior Management Sign-off

A package's percent fit to the required software functionality is very important, but people, not spreadsheets, should make software decisions. The final selection should come through a team vote. This facilitates more discussion and ownership in the decision and seals the recommendation.

Therefore, software fit and the pros and cons of technology, support, and vendor viability become back-up material. Of course, the overall team vote should closely correlate with this information, unless there is a major concern with a package that overrides any scoring system or list of advantages.

Each software module team gets one vote. In addition, the areas of technology, implementation support, software support, and financial viability each have one vote. Software functionality likely has more teams and, ultimately, has the highest impact on the decision. This is by design since if the functionality is a bad fit, a cheap package or strengths in technology, support, and other areas will not mean much in the end.

Accompanying each team's vote is a statement explaining the primary reasons they believe their recommendation is the best choice. If the vote comes down to a tie, the package choice is deferred to the evaluation manager and the steering team.

Obtaining senior management approval of the recommendation should not be like visiting the Wizard of Oz (the terrible man behind the curtain). In fact, there should be no big surprises if executives have received status updates from the team on a regular basis.

Prior to this time, vendor access to senior management was prohibited. It is now appropriate for the recommended vendor to give a presentation about their company and a high-level overview of important software functionality pertaining to your business. This provides the opportunity for management to get a warm and fuzzy feeling about the decision.

Management's job at this stage is to approve or disapprove the team recommendation. If they disapprove, they should send the team back to the drawing board to find a better package.

Managing Contract Negotiations

Never hand off the contract negotiations to the lawyers to fight out. The manager leading the software selection has been involved from the start and is in the best position to manage contract negotiations.

The goal is to keep the negotiations moving, to mitigate contract risk, but at the same time, to provide a voice of reason when negotiations get tenuous. Major revisions to draft agreements should be ironed out with the vendor sales team before getting any lawyers involved.

Software Licensing–Contract Tips

Vendors write legal agreements for one reason: To protect themselves. Initially, all contracts are heavily slanted in the vendor's favor. Thankfully, most ERP projects do not end up in court, but short of this, your organization still has plenty to lose.

First, most vendor promises, verbal or otherwise, should be incorporated into the contract, even when the vendor insists it is not necessary. If a dispute arises later, the vendor will have amnesia or those making the promises will have previously left the company. At this point, it becomes your word against theirs.

Everything is negotiable, but what you can expect to get depends on if you are a big fish (customer) in a little (software) pond or a little fish in a big pond. In any case, push hard for contract language limiting financial exposure, minimizing risk associated with software non-conformances, and anything that limits your flexibility.

Of course, be a tough negotiator, but also be a realist. While it is difficult to predict what a vendor may agree to, it is important to manage internal expectations on what one can achieve in negotiations. Do not be so unreasonable that a good vendor walks away from the deal. This does happen. For example, no vendor guarantees in a contract that the software will meet your business needs, no matter what their demo guy previously said.

When negotiating SaaS contracts or any externally hosted system, the negotiations include more than just items surrounding the application software. Many of the additional issues are traditional IT support considerations outside the scope of this book. However, many of the tips listed below relate directly or indirectly to SaaS. At the end of this chapter is a list of IT areas that one must ensure the SaaS vendor can provide.

Financial Exposure

When negotiating a contract one must protect against the worst-case scenarios because this is what legal contracts are all about. Worst case: the software is so bad you wish you never bought it, or the vendor is unresponsive to major software issues affecting your business. Most projects do not come down to this, but whoever has the money also has the advantage and limits their financial exposure.

When purchasing the software outright and implementing in multiple phases, some vendors allow phasing in the purchase at the same discounted price. Ideally, negotiate the best possible price for all software or licenses, and then cut separate purchase orders for what is required for each phase.

However, many vendors will not agree to separate purchase orders. The reason is the discounted price is based on volume (a single PO for all software in-scope) which is a firm commitment to purchase all of it. Without a purchase order in hand for all software, there is no commitment. Of course, this is why this approach limits the customer's financial exposure, so give it a try.

The next best approach is to purchase all software required at the discount, but to stretch out the payments. No doubt, a purchase order is legally binding, but if things turn ugly with the software, the vendor has more incentive to address the issues (because you have the software and most of the money). Of course, withholding payment is a last resort, but if the project heads to court, financial exposure is limited before there is a court ruling or a settlement.

It is important that the payment terms negotiated provide a reasonable time to discover major software bugs (non-conformances). On most projects, this is up to eight months (or longer if the software rollout is phased). A one-year payment plan, such as 50% down at contract signing and 25% after six months and the remaining 25% at one year, is probably the best to expect.

Licensing Cost

One should expect a discount off the list price for software licensing, but look for ways to get even deeper discounts. For examples, distinguish between regular users and causal users. A causal user accesses the system on an infrequent basis to look up information or run a report. It can be argued that their licensing cost should not be as much as a "heads down" daily user. Even when the vendor does not offer separate types of licenses that recognize this, many will provide an additional discount considering the number of causal users.

Next, is the question of future licenses. As suggested, license only the software necessary for the project, but extend pricing for potential future user licenses or additional modules. This accommodates unplanned users, near term business growth, or future projects. Get the price, not the same discount off the *then list price* (future list price), which will likely be higher. It is not hard to negotiate future pricing for up to eighteen months. Most vendors show flexibility here since they view sales outside the initial project scope as incremental.

Warranties and Remedies

Most vendors warrant their software free of defects for a limited time. A defect normally means a non-conformance to the intended design as represented by system documentation. If a software issue occurs within the warranty period, and the vendor states the software is *functioning as designed*, they do not consider it a defect.

Warranty periods initially offered are typically one to three months after the contract signing. Remedies or limits of liability accompany warranties. While all packages have software bugs that vendors address through regular updates, the contract concern is defects that become such an issue that they delay the project or the software is not usable.

First, make sure the warranty period is long enough to discover major defects. One to three months is not enough. The goal is to extend warranty expirations beyond the time any major defects would likely be discovered. On most projects within the first year some serious software testing has occurred, so eight to twelve months is a reasonable period for warranty expiration. Note that software modifications or enhancements performed in-house can void warranties. Of course, customizations made by the vendor on behalf of their customers should be covered under warranty terms.

It is important to address what happens if a major defect is discovered during the warranty period. Without some language beyond the standard terms, a required bug fix might become lost within the vendor organization.

There should be contract language requiring a written response from the vendor (within a specific number of days) regarding the disposition of a defect claim. The vendor should either acknowledge the issue as a defect or state the software is functioning as design.

When the vendor can prove the software is functioning as designed, usually the only recourse is submitting an enhancement request. However, even if their development group sees a need for the enhancement, the requested changes can take months or years to incorporate into the software. If it is a major issue affecting system usability, this is where stretching out payment terms as leverage may expedite the software change.

When the vendor acknowledges a defect, the contract should stipulate the number of days the vendor has to communicate when the bug will be fixed (such as in a previously scheduled software patch or new release).

Taking this a step further, the contract should require the vendor to provide a unique fix to your instances of the software if demonstrated that the problem is significantly affecting the ability to run your business. In this case, you should not have to wait for the fix in a future release.

When the vendor fails to meet the contractual requirements related to warranty, this is a major issue. In the end, if the software is not usable, the remedy is what can be negotiated in the contract such as refunds on items such as the software licenses, implementation costs, and maintenance fees paid to date.

Indemnification

Though rare, the contract should account for vendor responsibilities if accused of infringement by a third party. Indemnification clauses should state the company is held harmless of claims brought again the ERP vendor for infringements relating to the software.

First, infringement claims are typically not for the entire package but for specific components within it. At a minimum, the vendor should procure your rights to the software in question in such a way that allows for your continued use of the system. In the event this is not possible, the vendor agrees to replace the infringing software with an equivalent product at no charge, including implementation cost.

When the contract states the vendor has the option to cancel the agreement due to infringement and discontinue your company's use of the software, the company should receive refunds for software, implementation, and maintenance fees. This language is probably a mute point since the vendor is probably out of business, or close to it in this case. Again, this is an unlikely scenario, but it is best to cover your rights since the vendor is certainly covering theirs.

Source Code

Access to the software source code provides the ability to make modifications to programs and take complete control of the software if it later becomes necessary. Long-term, this is important if the vendor goes out of business and no other vendor purchases and supports the package. When a vendor does not provide their clients rights to the source code, always pay the relatively small fees to have it escrowed.

Software Support and Maintenance-Contract Tips

Software maintenance can be a separate contract and unpleasant surprises in this area are more common than with software licensing contracts. The list

below is the minimum level of support expected from any software maintenance agreement:

- Access to vendor support personnel (from both an application and technical standpoint) via phone and email
- Application or technical bug fixes (patches)
- New software functionality (new releases) at no charge
- Upgrades to keep the package compatible with new technology releases from third-party vendors (service packs)
- Access to the vendor support website for information such as application and technical documentation
- Access to customer support representatives for at least a 9am-5pm (eight hour) window every day

Support/Maintenance Fees

Many vendors offer a standard maintenance program while others provide additional options based on the level of support desired. When purchasing the software licenses outright, annual maintenance fees are quoted as a percentage of the total licensing cost.

For a given maintenance program, generally there is room to negotiate. Make sure you do because the list price for annual support is up to 25% of the software cost. At this price, one will pay for the software over again in just four years!

Whether negotiating maintenance fees as a percentage of the net software price or list price, the idea is to pay as little as possible. Where this goes, no one knows for sure—but depending on the package, negotiating annual main-tenance fees down to less than 15% of net price is not unheard of.

To keep the total cost of ownership as low as possible, try to limit annual fee increases to no more than the consumer price index. In addition, some vendors will wave the first year fees.

Longer-term maintenance contracts such as a three-year renewal cycle can yield higher discounts. But this might not be a good deal if all the money for the three-year period is due up front. Longer-term agreements can be worth it when the plan is to stay on the vendor support and if the deal allows for annual payments prorated over the contract period.

Right-size Support

All software vendors offer at least 9 to 5 support and most 24/7 coverage. In addition, some offer dedicated support personnel (by name) assigned to your account and preventive maintenance type programs. Of course, the more support, the higher the cost, and some services are overkill for many or can be obtained from a less expensive third-party vendor on an as required basis.

For example, most support agreements start upon contract signing. When the vendor refuses to forego the first year maintenance fees and offers several support options, consider when the system is to go-live. If the implementation timeline is more than twelve months, do you really need the highest level of support for the first year? The software is not in production and lower tier maintenance programs at least provide for bug fixes and answers to questions on a 9 to 5 basis. The year the software goes live, sign-up for a higher level of support (if necessary).

Also, I have yet to see where the premium price for dedicated support personnel is worth it in terms of the quality and timing of issue resolution. The vendor should be able to address problems in a timely fashion without requiring you to pay more.

In addition, for systems hosted internally, contract add-ons can crossover into support areas normally considered outside the scope of the ERP software and into routine IT system administration functions. For example, system tuning or monitoring can be performed by the IT group or periodically by a less expensive third-party vendor.

Finally, even with SaaS or any outsourced system, your ability to make application security or configuration changes provides more flexibility and reduces support costs. In fact, many consider this critical in order to respond quickly to business needs.

Maintenance Support Prior to Go-Live

Implementation consulting services (from the software vendor or a third-party firm) are different from software support services provide through the maintenance program. Nevertheless, make sure the maintenance covers the need for the vendor to respond to application setup issues or questions *during* the implementation phase.

Some contracts make no mention of this. But when a call is made to the vendor help desk with an application question, you might be told that support only covers software bugs or technical issues prior to system cutover (not application setup). Therefore, these questions must be referred to your application consultant. The vendor not only collects the software support fees, but maybe more consulting hours. The vendor support desk should at least attempt to debug a configuration issue or refer the client to the system documentation that answers the question.

Get the Software You Deserve

As long as support fees are paid, the company is entitled to new software releases at no additional charge for the existing software licensed. Of course, this excludes implementation costs or new technologies necessary to use the new release.

In order to get around this, many vendors market new functionality as a separate software component. Customers are later surprised to discover the new functionality must be purchased separately, since according to the vendor, it does not fall within the scope of the original software license.

No doubt, vendors usually have the upper hand here, since they can claim anything is a separate purchase. So, get a technical definition from the vendor of what constitutes separate software. My definition is separate programs using an entirely separate database, and *export/import programs* exist for interfaces. This is truly separate.

Review the vendor's development plans for the next few years and take nothing for granted. If new functionality is planned for the software to be purchased, and you need the capabilities, make sure it is included in a future release (free). If not, negotiate favorable pricing in the contract for a potential future purchase.

Returning Software or Licenses

The right to cancel support should be in all contracts, but short of this, ensure that the contract provides the ability to decrease the number of licenses or drop support for any module at any time. This provides the flexibility later to reduce unnecessary maintenance costs. If for no other reason, businesses change, and all the software and licenses required today may not be required in the future.

When reducing licensing, do not expect a refund on the original purchase. Nevertheless, when there are user licenses or software the company will never use, the associated maintenance costs can add up over time.

Beyond the right to reduce the number of licenses, software, or cancel the maintenance agreement entirely, there should be the ability to do any of these at *any* time. This includes a pro-rated refund considering the contract period and fees paid in advance.

For example, most standard maintenance contracts call for the automatic renewal of the current agreement unless the customer notifies the vendor of their changes at least 30 days prior to the end of the contract period. There is typically no pro-rated refund clause, and failure to notify 30 days prior results in paying for unused software licenses for yet another contract period!

Even with all the above in the maintenance agreement, some vendors after acquiring a software package from another vendor, try to ignore the original contract rules. In this case, the new vendor changes their maintenance policies and then attempts to make the change retroactive when a customer wants to reduce support. For example, after submitting a licensing change the new vendor threatens to reprice the entire maintenance contract at a higher price under the new rules.

To help protect again this situation, the contract language should specifically address this scenario. The maintenance agreement should remain in force even if the package is acquired by another vendor. If nothing else, the new vendor will acknowledge your foresight and take your claim much more seriously. Often times having contract language of any kind is better than having none at all, because many times it comes down to reaching a deal to resolve contract disputes. Perhaps in the dispute above the new vendor would agree to a one-time reduction in licensing without voiding the original contract.

SaaS-Additional Contract Considerations

Many fear that SaaS implies the loss of internal control of the system, including the ERP package. One thing is for sure: If there is a technical glitch in the software, your IT department cannot rush in to fix it! This is now in the hands of the vendor.

Cover these concerns and other technology support functions with specific contract language. Some may include service level or performance clauses.

This list includes data security, system back-ups, system performance, system availability, and disaster recovery to name a few. Additional items include data storage requirements, and hourly rates for software modifications (if applicable), configuration changes, and other types of change orders.

CHAPTER 6
SELECTING SOFTWARE CONSULTANTS

Seventeen Consultant Myths and Half-Truths

When many think of ERP success, they think of software consultants. At the same time, when many think of ERP failures and budget overruns, they also think of consultants (but not in such a favorable light).

Having spent the majority of my career in the shoes of a software consultant or an industry practitioner, I often wonder why many find it necessary to perpetuate the old myths and half-truths regarding the consultant/client relationship. Given the track record of ERP, we are not doing anyone any favors. We are causing more harm than good by setting the wrong expectations regarding what consultants are, what they are not, and what the organization should be doing.

My theory is these myths are self-perpetuating. Consultants want their clients to believe them (more billable hours), and their clients desperately hope they are true (even though deep down they know better).

Understanding the myths and half-truths can determine the level of consulting support required and the organization's responsibilities during the project.

Myth #1: Consultants will make us successful.

Truth: Consultants can educate, suggest, coach, and do the tough tasks, but cannot make their clients do much of anything. For most of the ERP *critical success factors*, consultants have no direct authority or control over the outcomes.

Myth #2: Consultants should take the project leadership role.

Truth: Whoever leads also owns, and consultants cannot own your project even if you insist they do. In addition, there is a big difference between *facilitating* a project versus *doing* it. On many projects, the consultants do everything, and the project team learns nothing.

This does not go unnoticed by employees. Many consulting firms grab and run with the project, and then wonder why employees are not committed, are disengaged, or resist the proposed changes. The employee attitude becomes "Great, go forth, and excel! We will point out the landmines once you hit them!" It is amazing how many do not understand this basic concept of change management.

Myth #3: Consultants are all-knowing.

Truth: Just because you pay someone $200/hr does not mean they know everything. Many naively assume all consultants are geniuses and later find out they are far from it.

There are many inexperienced consultants out there, but even the best ones will admit they do not know everything about the software. In addition, since every project is different, the application of the software tool is not identical in all situations. Even if the consultant understands many aspects of the system, it is unlikely that the consultant has experienced all potential usages.

Furthermore, no matter how much analysis consultants do, they will never know the business details and issues as well as your employees. When the project team is not fully engaged or not adequately trained on the software, the system design will miss the mark.

Myth #4: The more experts, the better.

Truth: Most projects do not need more experts—they need more solutions. It is not a good sign when, in the middle of a project, we suddenly need more consultants.

Too many consultants can crowd out or disempower the knowledge and experiences of employees that could play a larger role. Have you ever heard "No one is listening to my concerns" or "Joe seems to be clamming up"? When these statements come from creditable employees, this means that either there are too many consultants or they are not good listeners.

Myth #5: Consultants from expensive or prestigious firms have more insight.

Truth: The prestige of some firms might provide senior management with a good feeling and their consultants might be smart, but this does not mean they know anything about the software.

Myth #6: A consulting firm's track record with other clients says it all.

Truth: A vendor's record of accomplishment is important, but all firms have a few skeletons in the closet. At the end of the day, the only thing that matters is the knowledge and experience of the individual consultants working on your project—not someone else's.

Myth #7: Fixed price and progress payment engagements are risk free.

Truth: Anything that makes consultants more accountable is good, and there are plenty of ways to do this. However, if you can get great consultants for a very low fixed price (as seen on TV!), with few contract loopholes and a clear definition of their responsibilities and deliverables, hurry up and hire them. But good luck in finding such a deal (and watch out for the pitfalls discussed later).

Myth #8: Consultants with the best price and fastest implementation must know something that other firms do not.

Truth: Often, the opposite is true. An excessively low quote never helped any customer even when their management insists on a lowball schedule and price. It is best to hire consultants with the right skills and experience, who set realistic expectations with management and can deliver on their promises.

Myth #9: Implementation tools and templates make for better firms and consultants.

Truth: We definitely need tools since anything that makes life easier is helpful. The problem is most tools do not address project tasks that are the true bottlenecks.

For example, the intangibles of discovery, decision-making, and issue resolution drive the timeline, not how long it takes to physically set up the software. Also, any significant amount of custom software development such as modifications, data conversions, or interfaces is usually on the critical path. Tools and templates are not much help in speeding up these activities.

It is common when implementation tools are impractical to apply, full of bugs, or in the hands of consultants who do not know how to use them. This may even delay a project, or the tool becomes expensive *shelfware.*

The availability of more tools has affected hiring practices within many consulting firms. There is a misguided belief that one can hire smart people off the street and turn them into instant application consultants. Unfortunately, for their clients, it does not work that way.

Myth #10: A software consultant is also a good business analyst.

Truth: Understanding ERP software is one thing, but knowledge of industry practices and how to redesign business processes is another. The key is how the software is used in conjunction with redesigned workflows, new policies, roles, etc. Again, there is no such thing as a turnkey solution.

Myth #11: Consultants will transfer software knowledge to the project team

Truth: In spite of these promises, do not count on it. There are a dozen reasons why knowledge transfer may not occur and many relate to consultants.

Myth #12: We need more software consultants because we lack the internal knowledge and skills.

Truth: Maybe, but most organizations are capable of doing at least 30% of what software consultants are paid to do (much of which requires, little, if any knowledge of the system). With a knowledge transfer plan, most companies can do even more. This includes many of the software tasks traditionally performed by software consultants.

Myth #13: We need more software consultants because we cannot free up internal resources to participate.

Truth: Many consultants and managers too easily accept this as a rule; but if you look hard enough, get creative, and stop doing things within the company that add no value, plenty of internal resources may magically surface. One thing I discovered is when senior management really wants something to happen, they find ways to free up the right resources to make it happen.

Myth #14: Hiring new employees or training existing employees to fill application-consulting roles cannot be cost justified.

Truth: Not so fast. As mentioned before, the average time to implement ERP is one to three years. This is just the beginning since most companies will run their system another eight to thirteen years. Over this period, the cost of outside consulting support can really add up (if the goal is to support the system and users, and continue to leverage your software investment).

Myth #15: Once the project begins, the consultants will manage themselves.

Truth: Bad idea. Software consultants do not always operate with the correct assumptions about your business or make the right decisions, and sometimes they do things that are not in their client's best interest. Consultants must be managed.

Myth #16: Consulting firm management will be frank and honest with your senior management.

Truth: The good firms will be honest, but some do not want to deliver any bad news to senior management for fear of falling out of favor.

Myth #17: If the project fails, we can always blame the software consultants.

Truth: Yes, consultants are punching bags for some (and at times for good reasons). But when it comes to ERP disasters, internal folks rarely get off the hook easily. Even if lucky enough to escape the initial political fallout, do not forget the "walking wounded", whose careers are ruined.

Primary Service Provider

The primary implementation service provider is the firm selected for project management support and the majority of application consulting. This is the core of the consulting team, and they must be on the same page at all times.

On some projects, the primary vendor also manages other consulting companies and contractors involved. While all consultants must coordinate and play well together, the concept of *consulting firms managing other consulting firms* is never the best approach. Again, the internal project manager should manage all vendors.

As a general rule, the fewer firms and consultants to obtain the necessary expertise the better. Fewer vendors means fewer points of accountability, more consistent approaches, better communication, fewer hand-offs, and less finger pointing. This adds up to a project much easier to manage and with less built-in inefficiency.

Also, be aware that some firms use independent contractors to fill the need for application consultants. This is not necessarily a problem, but it is important to know the extent to which sub-contractors will be used.

A few independent application consultants are fine since many are well qualified. If you hire them directly (instead of through the primary firm) there are also cost advantages. Nevertheless, when the *majority* of the application consultants are sub-contractors, this amplifies all the risks noted above.

Another potential problem is that independents are more likely to be juggling other customer priorities outside the primary firm's or your ability to control, possibly leading to the consultant not being available as planned.

Hiring qualified software developers/contractors from other firms is usually of minimal risk and saves money. For example, cherry-picking good developers from "body shops" or other sources is usually not an issue when they are brought in at the right time and properly managed. Unlike application consultants, these resources play a more limited role and focus on completing specific programming assignments.

Types of Consulting Firms: Pros and Cons

There are three basic classifications of primary service providers: 1) The ERP software vendor, 2) A "Big Five" type firm, and 3) A third-party independent firm that specializes in the package. These are usually small to medium sized "boutique" shops.

Again, a few independent contractors can be used to augment the consulting team, but by definition, they are not the primary service provider.

Of course, each firm and consultant should be evaluated on an individual basis. However, based on my experiences as a software consultant and working with various software consulting companies over the years, my opinion of potential advantages and disadvantages of each type of firm are:

- **ERP Software Vendor Consultants**
  - **Pros:**
    - Single point of accountability for the ERP software and consulting services.

- Well-developed implementation methodologies, processes, and deliverables.
- More consistent level of application consulting expertise that tends to be better than average.
- Ease of consultant access to technical support resources within the vendor organization outside of formal support channels.
- More technical support resources.
- Dedicated training facilities, resources, and formal training programs.
- Potentially will make unique software customizations for their customers that later become part of the standard product offering.
  ○ **Cons:**
- Can be the only option available.
- Can be more expensive than third-party independent firms since the software vendor is often viewed as the safe choice.
- Project management consultants are prone to be "generalists," perhaps providing less *value add* in application or technical areas.

- ■ **Big Five Types**
  ○ **Pros:**
- More methods and tools for preparing consulting quotes, perhaps leading to a more realistic quote.
- Usually, firm management is more experienced to provide project oversight and guidance to the executive steering team and the project management team.
- Typically, additional areas of industry and process expertise within the firm (when truly needed).
- Well-developed implementation methodologies, processes, and deliverables.
  ○ **Cons:**
- Overall less experienced application consultants due to the common practice of hiring college graduates with no previous ERP software experience.
- Usually, much more expensive.

- Perhaps fewer technical resources available within the firm.
- Greater tendency to push additional services that are not required.

■ **Third-Party Independent Firms**
  ○ **Pros:**
  - Typically, software consulting for the package is the only goal of the company (do not consult for other packages or develop software-potentially diluting focus).
  - Perhaps the best value, considering their level of experience and hourly rates.
  - Project management consultants tend to be more hands-on since many have application consulting experience.
  ○ **Cons:**
  - Greater propensity to lowball quotes for competitive reasons.
  - Possibly less consistent use of implementation tools from project to project.
  - Fewer application consulting resources available (or heavy use of sub-contractors).

Local Versus Non-Local

Many place too much emphasis on the desire to have local consultants. Early in the consultant selection process, do not exclude any firm from consideration because of potential travel cost (unless the distance is very unreasonable).

The first criterion is the quality of consultants available, second is the hourly rate, and third is travel cost. When consultants are bad, it does not matter if they are local.

In addition, it is not always necessary for the consultant to be at your location to interact with the team or the system to complete important work. There are many ways to minimize travel costs; thus, local vs. non-local should only come into play when other factors are relatively equal.

The Consultant Selection Process

The search for a primary service provider should begin with the two package finalists. With some packages, there is not much of a choice other than the software vendor's consulting resources.

In general, a subset of those evaluating the software should also evaluate consultants. In addition, during the evaluation, no consulting firm should have access to senior management until the team makes its recommendation.

Evaluating the Actual Consultants

The consulting firm's track record does matter. But as said before, what really counts is the quality of the individual consultants that will work on your project. Selecting a consulting firm is a bottom-up evaluation starting with each consultant.

Beware of the pre-sales "window dressing", when certain consultants are brought in to display the expertise within the firm. They may know best practices and the software, but it might be the last time you see them—the old bait and switch routine.

All too often, many companies do not truly evaluate the qualifications or other attributes of the consultants proposed. They assume consultants are consultants and, more or less, all are created equal. But this is usually not the case. Not only should you evaluate their knowledge and experience, but their soft skills can make a difference in how effectively they work with clients.

First, request the resumes of proposed consultants and review them as you would a candidate for an important employee hire. Make sure the resumes are specific in terms of experience, skills, software modules, and client successes. Many resumes presented by consulting firms are not really resumes, but profiles that read like sales literature.

Next, do an old fashion interview and ask the tough questions. Make sure decision-makers (the project manager, IT manager, and those on the project team that will work directly with each consultant) are involved in the interviews.

The firm should have a list of every account the consultant has worked on over the past three years. Start making the phone calls, and when possible, speak with their project manager or functional analyst that worked directly with the consultant.

Finally, recognize that consulting firms propose consultants for new projects based on their current availability. When not completely satisfied with the consultants proposed, insist on better ones, and not just those currently on the bench. There might be better consultants immediately available or soon to come off another project. In the latter case, perhaps the firm can make them available much earlier than planned.

Evaluating Each Type of Consultant

There are five areas to evaluate when selecting the primary services provider including the *firm management* team, *project manager, application consultants, software developers,* and *technical support.* In this chapter, the focus is on the first three.

Someone within firm management, a partner or practice leader, should periodically attend Executive Steering Team meetings. The responsibility is to work with senior management as a peer to help guide the project down the right path.

Consulting firm management must be able to quickly gain the confidence of executives and have the ability to provide solid direction, education, and coaching at that level (and with the project management team). A good consulting firm can maintain positive relationships with senior management while also recognizing the customer is not always right.

The next role is that of the project management consultant. The organization is ultimately responsible for managing the project, but likely needs at least some project management support from the outside.

Look for these qualities in any project management consultants:

1. Has extensive project management experience with the ERP package selected.
2. Has managed at least three other projects of similar scope and complexity.
3. Has previous experience as an application consultant for the package selected, indicating ability to get into the details when necessary. The PM consultant must be more than just a meeting organizer.
4. Possesses knowledge and experience with the implementation methodology, including the purpose of each phase and deliverable. If they cannot communicate this at a sufficient level, they cannot manage the project.
5. Understands the key techniques to control the project. (see Chapter 19).
6. Demonstrates knowledge and experience with organizational change management strategies. Ask for the techniques or examples used on other projects.

7. Has experience with knowledge transfer strategies. Beyond formal project team training, what is necessary to ensure that software knowledge is transferred to the team during the project?
8. Possesses the ability to effectively coach and mentor the client project manager and steering team.
9. Can work effectively at all levels of the company (senior management, project team, functional managers, and end-users).
10. Sets realistic expectations with senior management.
11. Has good interpersonal and communication skills.

Each application team requires a person to fulfill the role of an application consultant. Look for these qualities when hiring an application consultant from the outside:

1. Has previous consulting experience with the assigned module on at least three other projects of similar scope, level of complexity, and industry.
2. Has business analysis skills.
3. Is able to provide direction and leadership.
4. Views the transfer of software knowledge to the client team as a priority and can explain how it will occur.
5. Is familiar with the firm's implementation methods and tools (or experienced enough to easily adapt).
6. Possesses soft skills including being a good listener, communicator, and coach. A consultant that does not listen may harbor preconceived solutions that are not the best for your business.
7. Fits the company culture. A software consultant should challenge the status quo, but not alienate people in the process.
8. Has some project management experience. An application consultant should know enough in this area to coach your project manager when necessary.
9. Has been certified on the ERP package, but actual experience is more important (certification is a plus).

A Real Methodology?
Any consultant who is experienced with installing ERP systems and under-

stands the package can work with any implementation methodology. But it is always best when the firm has a consistent set of methods, tools and deliverables they use on every project. Time-tested implementation processes improves efficiency and communication within the entire team.

During the sales process, most firms present PowerPoint slides with a few bubbles and arrows that depict their major software implementation phases, but for some this is about as far as it goes. Other firms sell their clients on the benefits of their implementation methodologies and tools, but then do not practice what they preach.

In either case, this is when the project management consultant later determines the *actual* implementation process. Depending on the experience of the consultant, the results can be somewhat unpredictable.

While there will always be some variation in approach, ask for examples of templates and deliverables used on other projects (with different project managers). These should be similar to the deliverables discussed in the next chapter. Also, have the firm demonstrate their tools, investigate these during reference checks, and ask each consultant if they have previous experience with the methods proposed.

Realistic Sales Quotes

Consulting proposals (quote, statements of work, engagement letter, etc.) contain estimates of project timelines and consulting costs. It is in your best interest to insist on realistic estimates since the organization has the most to lose if the numbers later prove to be severely underestimated.

Prior to selecting ERP software, we addressed the importance of documenting project-planning assumptions, risks, and constraints. Now is the time to refine this list since unlike software quotes, consulting bids are more subjective and driven by the experience of the person estimating. It is important that all firms bidding on the project have the same understanding of the project.

Estimates will always be just estimates, but perhaps equally important is that all planning assumptions should accompany all final quotes. No doubt, the consultants will ask questions when bidding and uncover more assumptions to add to your list. But the organization is responsible for ensuring assumptions are valid and the same for all vendors quoting.

In addition, always investigate major differences in the final quotes since the project timeline and consulting hours should be fairly close. For each consultant, estimated hours should be presented in weekly buckets within each phase over the entire project (i.e., phase/consulting role/weekly hours). This requires the firm to link consulting hours with the timeline, and the type of role and work each consultant will perform during each phase.

For application consulting estimates, make sure the quote reflects their involvement in new software development. Application consultants are required to write programming specifications and test any type of custom development. Often, application consulting quotes fall well short of the time required in these areas.

The Fixed Price Game

Most consulting firms avoid fixed price and progress payment projects for valid reasons. First, most prefer not to get into the exhaustive contract wrangling required for these deals.

Second, once the project is complete, there is a good chance the software functionality delivered and the implementation cost will not meet their client's original expectations. Finally, most firms do not want to assume the risks associated with a client that does not deliver on their responsibilities.

Even when consultants quote a fixed price, it must be realized that most are only fixed until further notice. This is because most firms are better than their clients are at writing software consulting contracts. No matter how well you dot the i's and cross the t's, the verbiage is always open for interpretation, and debate, and scope creep is inevitable. The result is expensive *change orders*.

When it is possible to lock a firm into a price that is truly fixed, keep in mind that consultants are in business to make money. The fixed price probably has plenty of fluff, and you will not get a refund if the project requires *less* consulting hours.

The "not to exceed" clause adds little comfort. One thing is for sure, consulting hours will expand to fill all available space. When this occurs, you might have been better off going time and material.

In addition, most consultants are very good at making software function to meet the bare bones language of any agreement. There is a high probability that the functionality will be delivered, but not to the level required by the business.

Finally, ERP software is unforgiving. When we rush or cut corners to meet arbitrary dates or budgets, there is a price to pay later.

Of course, this is not to say fixed price or progress payments never make sense. They are tools in the toolbox but tend to apply more to projects of very limited scope. In no case are these agreements a substitute for taking ownership of your project, nor do they reduce the need to manage your consultants.

Implementation Services – Contract Tips

The following are a few additional tips when negotiating the consulting services agreement:

1. ***Firm partners and practice leaders are non-billable.***

    As mentioned, a partner or senior manager of the firm should provide periodic guidance to the executive steering team and project manager, not just make grand appearances and soak the customer for more money (their rates are much higher).

    In fact, the time of a firm partner or practice manager should not be billable for this limited time spent on the project. One reason is some already receive bonuses or a percentage of billable hours generated from the account. Also, they are executives of the firm. Unlike application consultants, for example, firm management has a major stake in project success that goes beyond billable hours. Once the word gets out of a project failure, this can hit a firm partner right in the pocket book in terms of future sales.

2. ***Do not pay for consultant schedule changes.***

    Some vendors want a commitment for a certain number of hours per week for each consultant to ensure availability. They next want to charge you for schedule changes. Never accept this! If their consultants are not available, do not hire the firm in the first place.

    Alternatively, work with the firm to develop a schedule for consultants based on your needs. Again, each consultant should have an estimated number of hours per week over the entire project as a starting point. However, the number of hours provided each consultant will vary by project phase and *will change* based on *actual* project needs. Of course, there should be an attempt to minimize changes once the schedule is firmed up and provided to each consultant, but it does happen. In fact, the consultants will probably change the schedule more than you will!

3. ***Do not pay for consultant learning curves.***
For example, one may hire a strong application consultant who is not familiar with a new implementation tool to be used on the project. This is not necessarily a problem, if the company gets a reduced rate while the consultant is learning it.

   In addition, *tag-along consultants* (shadowing other consultants for the sake of learning) are fine too, but the hourly rate for the primary consultant should be lower for a period to recognize the time spent teaching the other consultant. Also, the trainee consultant should not be billable when in learning mode. When a tag-along consultant is productive, their rate should be lower to reflect less experience.

4. ***Expect accurate time reporting.***
We want accurate billing, whether this means more hours billed or less. Consultants should bill by rounding their time to the nearest tenth of an hour. Many round up to the nearest quarter hour. This may not seem like a big deal, but do the math. If four consultants for twelve months are billing just one extra hour a week because of rounding, this will cost the company at least $35,000.

5. ***Off-site work must be pre-approved.***
A project like ERP cannot be successful if the consultant is mostly working *long distance.* The majority of consulting time should be on-site to enhance focus and communication. Of course, there are times when a consultant does not need to be on-site to be productive, so take advantage of these opportunities when they arise to reduce travel cost. Any work to be performed off-site should be approved in advance by the client project manager. We want to void invoice surprises.

6. ***Define what else is non-billable.***
In addition to the items previously mentioned, be specific in the

contract regarding other consultant activities that are not billable. Non-billable time should include travel time, breaks/lunches, and any communications with other clients while supposedly working on your project (watch out for this one).

7. ***Place reasonable caps on travel expenses.***
Place a reasonable limit on travel and living cost including airfare, hotels, meals, and rental cars. The project manager must approve any deviations to these limits in advance. In addition, negotiate discount rates at a good hotel nearby, to reduce travel expense.

8. ***Do not agree to the administrative fees***
Never agree to pay an invoice mark-up to cover the firm's "administrative time and supplies costs" while working on the project. This is the firm's cost of doing business, and their consultants will probably use mostly your materials and supplies anyway!

9. ***Retain the right to terminate without termination charges.***
The company should have the right to terminate any consultant at any time and for any reason with no termination charges. A few weeks advanced notice is reasonable.

10. ***Additional considerations for fixed price or progress payments.***
On any project, the sooner we get into the project details the better information we have to make decisions. But when a fixed price or progress payment agreement is the right choice, go much deeper into the project assumptions *earlier* and incorporate them *in the contract*. This means more analysis and consensus on all the project planning deliverables before signing any contract.

This is not to say one should shortcut these items for time and material projects. In either case, we want good estimates to avoid major surprises. But on time and material contracts, it is acknowledged by all parties that the true time and cost "is what it is."

On fixed price or progress payment projects you will get something for the price or progress, but it is important to understand what *something* means. It is all about what is to be delivered for the cost incurred.

Therefore, pay special attention to all areas of the project scope, especially the list of software features and functions and any custom software development included. Equally important, are the list of deliverables for each phase, what each should look like, and who is primarily responsible for each (consultants or the company).

CHAPTER 7

THE ERP ROADMAP

The Sales Proposal Is Not "The Plan"

Many view the sales proposal from the software consultants selected as the project plan (scope, resources, schedule, budget, etc). However, most statements of work, cost estimates, and schedules developed by consultants at this stage are far from reality.

The acceptance of the sales proposal is nothing more than the end of the consultant selection phase—not something we can use to manage the project. In all fairness to consultants, there is usually not enough information or justification for the kind of resources necessary to develop a valid plan when bidding.

Also, vendors do not make money on proposals unless accepted by the potential client, so they spend most of their time trying to get their client to accept the proposal, and less time making sure it is right.

From the organization's perspective, the primary purpose of the quote should be to select the most qualified consultants, fine-tune project-planning assumptions, and get decent estimates from all firms bidding to provide input into your own project estimating. The real project plan is developed when the project organization is finalized and when the consultants who will work on the project are on-board.

Owning The Plan

Anyone that has been around ERP long enough understands that meaningful involvement of key employees is critical. Therefore, it is amazing how many projects start with consultants behind closed doors, developing a detail project plan in a vacuum. They later unveil it as some artful piece of work and present it to senior management for sign-off.

Keep in mind that *blessing* a plan is very different from owning it. When consultants own the plan, ownership is in the wrong place. This issue is often reflected in management's message to software consultants: *A great plan, now hurry up and make it happen.*

Interestingly enough, even some of the best software consultants are guilty of planning in a vacuum. There are several problems with this approach.

First, it prevents the rest of the project team from getting heavily involved from the beginning. If nothing else, at least get the project started on the right foot.

Second, if the project manager and steering team are to understand the plan, they must have more than a token role in developing it. When they get their hands a little dirty, there is a much better appreciation for what the project entails. The role of the project management consultant during this time is to provide knowledge of planning, guidance, and tools, as well as to ensure the plan is valid.

Third, no matter how much analysis consultants do, they are never aware of the subtle aspects of the company that can have a major impact on the validity of the plan. In addition, failure to seek input from the team leads, functional analyst, IT manager, application consultants, and other stakeholders outside the team, is a missed opportunity to build a better plan. It also increases the likelihood that these same people, who must execute the plan, will not support it.

Get the Help You Need

Planning is not a place to cut corners. If you need more education and training in project planning, go get it. If you need more project management consulting to develop a valid plan, spend the money. However, if the existing project management consultant does not add value in this area, he will probably not add value anywhere else.

When developing a plan, it is fine to use various types of planning templates as a starting point. Examples might include project charters, detail schedules, a list of responsibilities, budget spreadsheets, scope items, etc. However, not tailoring the plan to the specifics of the project, results in much verbiage and formality, with little substance.

Avoiding the Icebergs

In many ways, managing an ERP project is like steering a large ship. Turning the wheel does not mean the ship responds instantaneously. If the Captain has not anticipated what lie ahead, by the time the wheel is turned, it could be too late. You are now aboard the Titanic! In terms of ERP, this means delays, rework, and cost overruns simply due to poor planning. As mentioned, ERP projects are never executed flawlessly, but good planning offsets less than perfect execution every time.

The ERP roadmap at the highest level includes the major phases of project *preparation, planning,* and *implementation.* The steps to prepare for the project were covered in previous chapters.

In the chapters that follow, the specifics of the planning and implementation phases are addressed. The purpose of this chapter is to understand the roadmap, the phases and deliverables, and the dynamics within each phase. The steps to implement ERP are not as serial as most flow charts suggest since there are many interactions between phases and deliverables. Understanding these interactions is important.

Planning Phase

No project plan can anticipate all the twists and turns, but many are abandoned soon after the project gets started. There is a difference between a plan that can be used as a handle to manage the project (as imperfect as it is) versus one that exists only for the sake of posterity. In the latter case, we are operating in the blind and later make dumb decisions to catch up to a date that was never agreed upon or valid to begin with.

The formal planning phase expands the scope and depth of previous plans, preparations, and assumptions. The outcome is the baseline plan. This is intended to layout the project detail and be used as tool to measure progress.

The complete set of planning deliverables has different names within the industry, such as the project charter, scoping and planning document, the master plan, etc. No matter what it is called or the format used, the overall content should be the same.

The project manager and project management consultant drive the planning process, facilitate the involvement of others, and create the actual deliverables.

The planning phase runs somewhat concurrently with other major activities. This is no accident: The preliminary analysis by the application consultants serves as additional input to the plan, as well as team software training and current process analysis. The more the team understands about the software and processes, and the more the application consultants understand the business, the better input they can provide.

Once some of the earlier planning steps are completed, a pre-kickoff meeting is held with the team. The agenda is to introduce the consultants, review the project organization, announce the team software-training schedule, and launch the *As-Is* process analysis. Usually by this time, the rollout of the systems infrastructure to support the project has started.

Once the steering team has approved the baseline plan, the project is officially launched with a kick-off meeting. This meeting involves the entire steering team, project manager, application teams, technical teams, consultants (firm partner, project manager and application consultants), and stakeholders recognized in the project organization chart. The executive sponsor chairs the meeting, and the project manager (not the project management consultant) presents the final planning deliverables.

Figure 1 on the next page represents the planning phase including the relative sequence, interrelationships, and timing of each deliverable.

Implementation Phases

The implementation phase is depicted in Figure 2 at its highest level. Unless using the ill-advised *load and go* approach, all ERP projects progress through these stages to some degree. Therefore, the following is a relevant discussion regardless of the specifics of the implementation methodology.

Discovery–Defining the Right Questions

With any project the scope of ERP, initially there are many unknowns. The sooner we can move the project from a non-deterministic to a more *deterministic* state, the sooner we can make informed software and business process decisions.

The primary goal of the discovery phase is to begin cresting the learning curves associated with the software and business issues (before the design phase). Everything about the project *will* be discovered at some point. But the choice is to discover as much as possible now versus later. When this phase is cut short and we jump right into design or construction of the system, unpleasant surprises progressively cost more to address.

This phase consists of four major components that interact closely with one

Figure 1 – Project Planning Phases

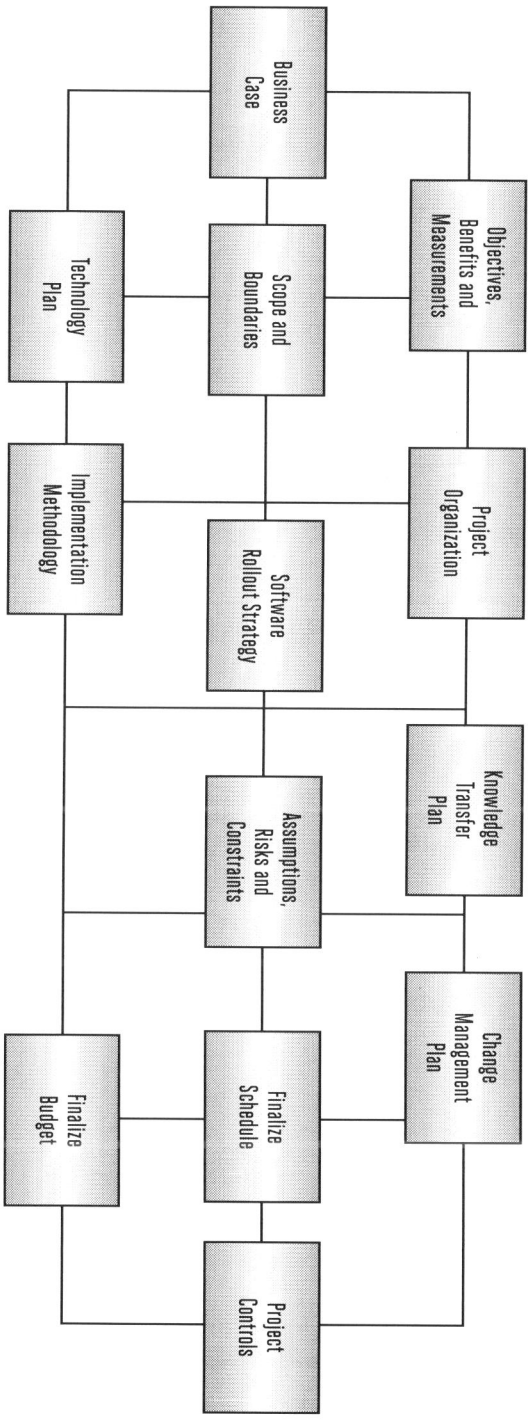

another. These include the *As-Is* processes (what we do today), *improvement opportunities* (what could be better), and *project team software training* (software capabilities). These three tasks feed the fourth component: the software *prototype/modeling* phase.

Prototyping is often misunderstood. It is not part of the design phase, rather it is the heart of discovery. The concept of prototyping has existed in engineering disciplines for centuries. Within the context of ERP, the idea is to quickly build a model (mock-up) system in a test environment, apply the software to business processes and needs, and experience the results first hand.

Before the prototyping begins, there has been some degree of requirements definition. Requirements are necessary to evaluate ERP software. The *As-Is* analysis yields more insight into business needs. These up-front requirements gathering processes are necessary.

However, prototyping takes the definition of software needs one step further. It reshapes, redefines, and accelerates the definition of requirements and the understanding of the software capabilities and set up. This is the *solutions driven* element to requirements definition. That is, the capabilities available in the software serve as examples of what could be and spark a much deeper conversation of business needs that is not possible prior to this point.

Prototyping also begins to expose the gaps between the vision of what the company wants do in the future versus what the software will allow it to do. Software limitations may adjust expectations, procedures, or could potentially lead to software modification. All of the outputs of discovery are the primary inputs into the design phase.

Design–A Stake in the Ground

When the ERP system goes live, it is similar to landing a big airplane in rough weather with 300 passengers on-broad. The first goal is to bring it in on the runway, keep it there, and handle the speed bumps once on the ground. This will not be possible without addressing the majority of the system design considerations during the *design* phase.

This phase determines how the software package will be set up and used within the business, as well as the specifics of any custom programming. At the end of this phase, will software requirements, new processes, and the system design be 100% complete? The correct answer is no because the steps

Figure 2 – Implementation Phase

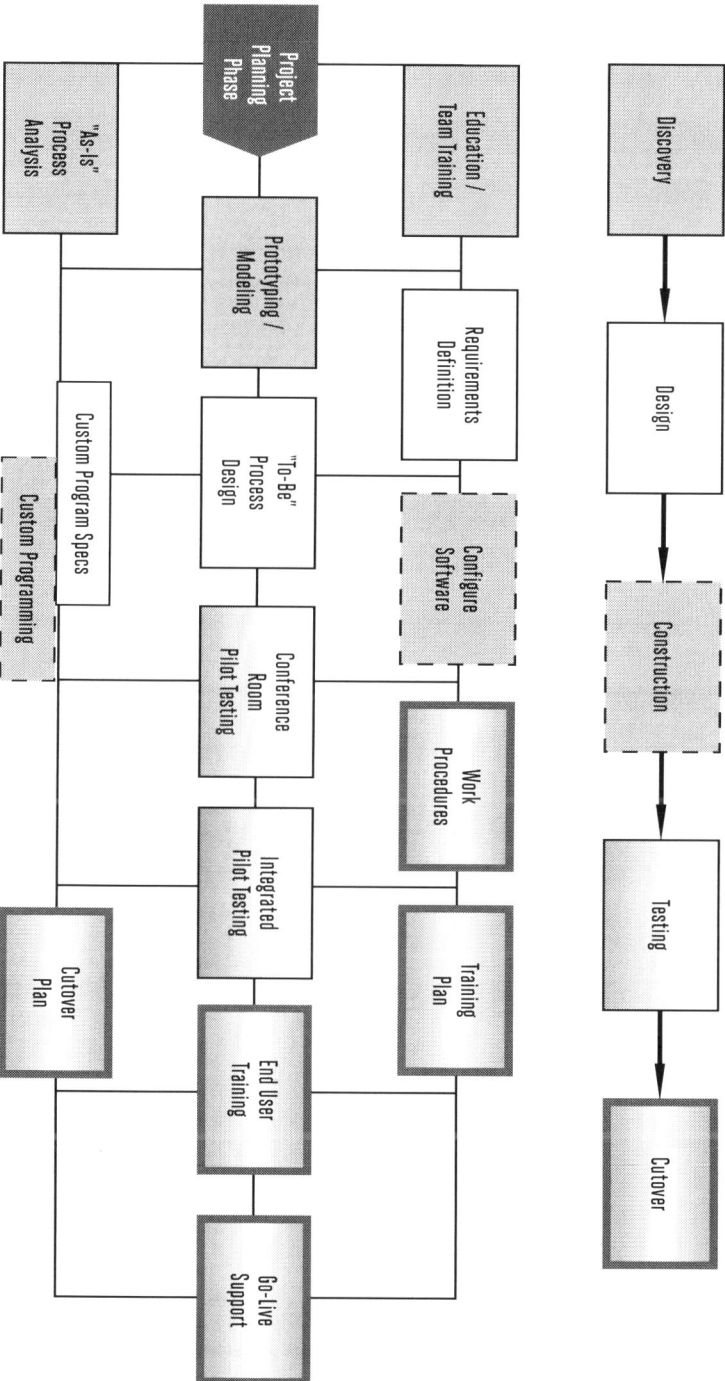

and information required to implement any system are never serial in nature. But the design phase is the first point at which we build a much higher level of consensus on the use of the software within the business.

The design phase involves brainstorming and hands-on informal testing by the team, as well as feedback and validation by stakeholders. Preliminary design concepts will come and go, and at times, the team will take two steps forward and then three steps back. These dynamics lead to a better design and ultimately a better system.

The major outputs of the design phase include new business process maps (*To-Be*) and specifications for writing custom programs (data conversions, interfaces, software modifications, etc—perhaps some of the biggest challenges).

Once identified, custom software development tasks are prioritized based on criticality, level of effort, and when first required during the formal testing cycle.

Construction–Building the Beast

The construction phase consists of building the system to the design blueprint (in the test environment) in preparation for formal testing. It includes configuring the software to support the new process designs, documenting the configuration, and custom software programming.

Preliminary software set up starts with prototyping and evolves through design. During construction the configuration is adjusted and expanded to cover the scope of the system. Once the configuration is complete, part of the construction phase is to begin the system setup documentation.

Note the custom programming portion of the construction phase to a degree runs concurrently with the test phase. Usually, it is not advantageous to delay all testing until all programs are written.

Testing–Six Types

The test phase should be highly organized, structured, and comprehensive. This is what separates formal testing from informal testing. The earlier informal "testing" was really about learning, validating certain software capabilities, investigating design concepts, or limited testing of certain programs.

Testing is more than identifying and fixing software bugs. It always results in adjustments to previous designs and software setup, and provides input into finalizing work procedures, policies, and training materials. It is also the project team's last opportunity for learning prior to the system cutover.

There are six major test cycles discussed at length in Chapter 18. These include the *Conference Room Pilot (CRP), Integrated Conference Room Pilot (ICRP), Limited Parallel Pilot, User Acceptance Testing, Volume/Stress Testing, and System Cutover Testing.*

Cutover–Pulling It All Together

Once the software is ready, cutover includes the activities to prepare the organization for using the system and the steps required to move the system from the test environment into production. The major tasks include:

- **Detail work procedures** describing (step-by-step) how to do specific job functions using the software and the key policies that should govern how this work is performed. The writing of work procedures should start toward the end of testing. When started prior to this time, this documentation will require constant revision. Some consider the development of work procedures as part of testing. I disagree. The purpose of testing is to test, and writing procedures at the same time dilutes this focus.

- **End-User Training** includes preparation of training facilities, equipment, schedules, lesson plans, materials, training data, and conducting the actual training.

- **The System Conversion Plan** addresses all system migration and conversion steps, the sequence, specific dates and times, responsibilities, and verification to ensure each step occurred as planned. One of the most avoidable problems at go-live are issues from a poorly planned or executed cutover. Therefore, carefully plan it and perform a dry run prior to the actual event. Finally, always have a strategy to rollback to the old system in case the cutover goes awry.

  If the thought is to perform a *Go-Live Readiness Assessment,* this is the time to do it. But as said before, if all has gone well up to this point, the additional expense for an assessment is usually not worth it. When considered necessary, hire a software consulting firm other than the one currently working on the project. The main goal is to have a separate and unbiased set of eyes review the implementation steps taken, deliverables, software configuration, procedures, training materials and conversion plan.

- **Post Go-Live Support Plan.** The team should develop a plan to support the end-users immediately after go-live. For several weeks after the initial cutover, the power users and functional analysts should be physically located in the work area to answer questions, address issues, or escalate issues for resolution. The application consultant should be readily available to address questions and issues that the power users and functional analyst are unable to address.

    This question/issue resolution support structure not only helps the users, but defuses issues and user frustrations that may otherwise get blown out of portion (causing the project a lot of unnecessary, and perhaps untrue, negative publicity). Also, after several weeks there should be short follow-up training sessions.

    There also should be a long-term support plan that begins after the immediate go-live issues are resolved. This plan should be developed by the project manager and steering team. This may include organizational changes to put the support people in the IT department or somewhere equivalent.

Deliverables-A Project Manager's Best Friend

The ERP road map is necessary for planning and navigation purposes, but when it comes to execution every project lives or dies by deliverables. The only way to successfully complete an ERP project is to eat the elephant one bite (each deliverable) at a time.

Deliverables represent three things. First, they are the sequential building blocks of the project, with each representing a stage of project completion. Therefore, deliverables are used to schedule, assign accountability, and measure progress. In fact, the project master schedule is nothing more than the list of deliverables, dates, and responsibilities. Of course, it is best to have a detail schedule to support the overall timeline!

The definition of roles and responsibilities, though important, does not go far enough in terms of desired outcomes. A review of deliverable formats (before the work begins) clarifies responsibilities of the team and establishes a higher level of accountability. In other words, here is exactly what is required.

Third, deliverables improve communication. On most ERP projects there are some unavoidable hand-offs when the work of one team serves as input into tasks another team must perform. In addition, deliverables enhance communication with stakeholders outside the project team.

The concept of deliverables is not intended to eliminate the need for other forms of communication. Collaboration is necessary to avoid the *throw it over the wall* mentality that is so often seen on large projects.

Finally, deliverables are also checkpoints for quality control. "Quality" means ensuring that deliverable expectations of stakeholders are met. This can be accomplished without creating a bureaucracy such as separate quality analyst, highly stringent deliverable formats, or formal sign-off.

Deliverable documentation examples are not hard to find so there is no need to re-invent the wheel. But some deliverables are not as tangible as a completed document for a final team review. Therefore, the project manager must find ways to make the outcome more tangible and measurable. For example, how do we know if a particular team learned anything during the initial software training? Upon returning from training, a brief software demonstration conducted for the benefit of other team members is a way to make this *learning* deliverable more tangible.

What About Software Upgrades?

In many respects, a system upgrade to your existing ERP product is not much different than a new implementation when major software functionality changes are involved. The biggest difference is there may be software modifications to replace with standard functionality or to retrofit into a new release. But in either case, the project should pass through the same implementation doors. You still have to plan it, discover it, design it, test it, etc.

Project Deliverables

Unfortunately, within the ERP industry there is no standard glossary of terms for project phases and deliverables. Many vendors add to the confusion with their own proprietary jargon that changes week to week.

Besides, I am not sure if terminology is relevant as long as the project has the right deliverables, the team understands them, and responsibility is assigned. Below is a list of deliverables for the implementation phase and the important project management questions each addresses.

Figure 3 – Planning Deliverables

KEY DELIVERABLES	PROJECT QUESTIONS ANSWERED:
Business Case	- Why is the project necessary from sr. mgmt perspective?
Objectives	- What is the project intended to accomplish?
Benefits	- What tangible and intangible project benefits are expected? - Who on the steering team is the champion of each benefit? - What are the critical success factors for realizing benefits?
Measurements	- How will we measure for success and sustain the desired behaviors?
Methodology	- What steps are required to implement the new system? - What is the content and format for each deliverable? - What deliverables are the responsibility of the organization? - What deliverables are the responsibility of the consultants?
Scope and Boundaries	- What is included in the project (and what is not included)? - Are all scope items / dimensions documented?
Software Rollout Strategy	- What areas of the business are to be implemented and when? - What degree of process standardization is expected?
Project Responsibilities	- What are the project teams and their relationships? - What are the expectations of each team and project role?
Project Team	- Who is on the project team and in what role? - How much time will each team member commit to the project? - Are there enough resources?
Knowledge Transfer Plan	- What type of knowledge will be transferred and to whom? - How will the team acquire knowledge throughout the project?
Current IT Environment	- What systems, technologies and interfaces exists today? - What are the system's major functions and user groups?
Technology Plan	- What technologies are required for the new system? - When will the technologies be installed? - What are the IT staffing requirements and training plans?
Master Schedule	- When will the project, each phase and deliverable be complete? - Is the schedule realistic / achievable?

Figure 3 – Planning Deliverables (*continued*)

KEY DELIVERABLES	PROJECT QUESTIONS ANSWERED:
Detail Schedule	- What tasks must occur to support the master schedule? - When will each task be complete and who is responsible? - What tasks are on the project critical path?
Budget	- What are the budget items and are all expenditures included? - How many hours are budgeted for each consultant per week?
Assumptions / Constraints	- What assumptions shape the schedule, budget and benefits?
Risks	- What events could negatively effect the project? - What are the contingency plans should these occur?
Change Mgmt Strategy	- How will key stakeholders be involved with the project and when? - What should be communicated to each group of employees? - Who is responsible for the communication and when? - Is the executive sponsor involved in these communications? - Who are the potential naysayers?
Project Controls	- What methods will be used to keep the project on-track?
"Official" Project Kick-Off	- Does the executive sponsor play a major role in the kick-off? - Are all project teams and key stakeholders present?

Figure 4 – Discovery Deliverables

KEY DELIVERABLES	PROJECT QUESTIONS ANSWERED:
"As-Is" Process Maps	- Are the meetings cross-functional (involving the right departments)? - Are the right employees involved (including those that do the work)? - Do the maps represent how the work is actually performed? - Do the maps capture all elements of each process (roles, systems, etc.)? - How do we conduct business today?
Problem Analysis	- What are the major issues within each process? - What are the root causes? - How is each problem classified (procedural, policies, systems, etc.)?
Improvement Opportunities	- What are the team's ideas for improving the processes? - What are the low-hanging fruit?
Education and Training	- Has the executive steering team been educated on ERP? - Is best practice education required for the project team? - What ERP software training is required for each team? - Will the software training include how to configure the software? - Who is to attend each training course and when? - Is additional skills or tools training required for the team?

Figure 4 – Discovery Deliverables (*continued*)

Initial Software Set Up	- What are the steps to configure the ERP software package? - What configuration settings are applicable to the business? - What is the "first-cut" software setup to support prototyping?
Prototyping / Modeling	- What was learned about the software during prototyping? - How well does the software address our major business processes?

Figure 5 – Design Deliverables

KEY DELIVERABLES	PROJECT QUESTIONS ANSWERED:
Key System Requirements	- What is the vision for conducting business in the future in each area? - What are the important software needs within each module?
Software Demonstrations	- What do key stakeholders and end-users think about the software?
Design Reviews	- How does the project team propose we use the system in each area? - What are stakeholder concerns about the proposed designs?
To-Be Process Maps	- How will the new business processes function using the software? - Are the project team and stakeholders in consensus on the *To-Be*? - If not in agreement, what are the major issues/concerns?
Software Gap Analysis	- What software limitations were discovered during design? - How will each limitation be addressed (procedures, software mod, etc.)?
Major Business Changes	- What key business changes are necessary within the *To-Be*? - Does the steering team support these business changes?
Software Development List (data conversions, interfaces, reports, modifications)	- What software mods have been approved by the steering team? - What are the development tasks, priorities and completion dates? - Who is responsible for programming and testing each program?
Data Conversion Design	- What are the requirements for each data conversion program? - Are there any major data mapping issues between systems? - What data conversion tools are available within the ERP package? - What data conversion programs are required early in testing? - Is data clean up required and who is responsible?
Interface Design	- What are the requirements for each interface program? - What interface tools are available within the ERP package? - Are software mods required to support the interfaces?
Software Mod Design	- What are the requirements for each software modification? - How will each modification be accomplished in the system? - Will the proposed designs minimize retrofits for new releases?
Reports Design	- What are the important reports identified so far? - What are the requirements for each report?

Figure 6 – Construction and Test Deliverables

KEY DELIVERABLES	PROJECT QUESTIONS ANSWERED:
CONSTRUCTION PHASE	
Baseline Software Set Up	- What software set up is required to support the *To-Be* Processes? - Are initial menus set up? - Has the software been unit tested to fix obvious setup issues?
Setup Documentation	- How is the software set up / configured ? - What are the business reasons for selecting each setup option? - Will the client team write this documentation?
Software Development (data conversions, interfaces, reports, modifications)	- Are there enough software development / programming resources? - What are the specifications for writing each program? - Are stakeholders in agreement with how the programs will work? - Will each program be unit tested (by the developer and application team)?
TEST PHASE	
CRP and ICRP Test Plan	- What processes are to be tested, who will test each and when? - Will business data be loaded manually for the 1st round of CRP?
Test Environment	- Except for setup parameters, is there a clean database to start? - Is the test system representative of the future system (in production)? - What data is to be loaded to support testing (given the test cases below)?
Test Cases	- What scenarios will be tested within each business processes? - Are custom programs slated for plenty of test cases? - Do test scenarios cover 95% of what could occur in each process? - Is the team (not consultants) defining the great majority of test cases? - Is the team (not consultants) performing the great majority of testing? - Are test cases more complex with each round of testing?
Test Results / Fixes	- Are all test cases dispositioned as "pass or fail" by the team? - Are issues being documented in the issue list? - Once a program issue is fixed, are other related programs retested? - Have all setup changes been made prior to the next round of testing? - What was learned from a policy, procedure, and training standpoint?
Volume Test	- How will the volume test be conducted? - During this test, were there any system performance concerns? - How will system performance issues be addressed?
Parallel Pilot Test	- Is the pilot area a good representation of the overall software scope? - What major issues were discovered during the pilot?
User Acceptance Test	- Do key stakeholders/end-users concur the system is ready for go-live?

Figure 7 – Cutover and Post Go-Live Deliverables

KEY DELIVERABLES	PROJECT QUESTIONS ANSWERED:
CUTOVER PHASE	
Work Procedures / SOP	- How will each group of employees perform their jobs within the system? - Beyond the software steps, are key policies and procedures in the SOP?
End-User Training Plan	- What groups must be trained in each application area? - What courses are required in each application area? - Will the functional analyst be the instructor and the power users assist? - Who will attend each course? - Are there enough facilities? - What is the lesson plan for each course? - What overview materials and training exercises are required? - What data must be loaded in advance to support each training exercise? - What is the training schedule? - Did the executive sponsor communicate the training schedule?
Training Environment	- Are the training facilities and equipment ready? - Is the training system ready and has it been tested? - Does the training system look exactly like the future production system?
Conduct End-User Training	- How did end-user training go? Who failed to attend?
Conversion Plan	- What specific steps are necessary to cutover to the new system? - What is the exact timing of each step and who is responsible? - How much time is needed for the conversion and when will it start? - Has the cutover plan been tested?
Go-Live Support Plan	- What team members and power users will support each area at go-live? - Who on the team will be physically in each work area at go-live? - Are all shifts covered? - What is the process for escalating issue resolution within the team?
Production Environment	- Have all programs been successfully moved to the production system? - Are all menus, security, and hardware in place and tested? - Have batch reports been set up in the job scheduler and tested? - Have all IT procedures been defined? - Have all conversions steps been completed and verified in production?
Go-Live	- Do it and hope it works.
POST GO-LIVE PHASE	
Post Go-Live Support	- What issues are user experiencing and what must be done to fix them? - What are the users questions and follow-up training required?
Long Term Support Plan	- Who will maintain the work procedure / SOP documentation? - Who will training new employees or those that change jobs? - Who will make software setup changes to support the business? - What organizational changes are necessary to provide this support?

CHAPTER 8

BUSINESS CASE, OBJECTIVES, AND BENEFITS

Clearing the Path

The initial steps in developing a project plan are establishing the business case, objectives, and benefits associated with the project. This analysis should begin before selecting software and is finalized during the project planning phase.

The business case is the justification for doing the project. It establishes how the project supports the overall goals and strategies of the business. Stated in business terms, it answers the basic question every employee will have, "Why is the project important to the success of the organization and, therefore, important to me?"

ERP represents change, in not only the computer systems, but also how we run the business and how employees perform their daily jobs. Therefore, senior management must be heavily involved in developing and communicating the business case. They are the only ones that can truly own the justification for the change.

If nothing else, the business case is about supporting the project team. They are the change agents—the foot soldiers on the frontline. Management is asking them to go beyond the call of duty.

In order to clear the path for the team, management must elevate the importance of the project above the agenda of any department or individual. If not, the team will face unnecessary resistance and the need to constantly defend the validity of the project. Of course, the team must help sell the system to employees, but they cannot justify it.

The Compelling Need for Action

The more compelling the business case, the more urgent the change imperative. There are many compelling reasons to implement a new ERP system. Many are justified based on new business strategies or challenges, but some build upon existing strengths. For example, if short product delivery lead-time is currently a competitive advantage, examine how ERP might further protect this advantage.

Communicating a compelling business case is not just what you say, but how you say it. The following are a few *less-than-compelling* business cases:

- **Our current systems must be replaced because they are old.**
  - While this may be one reason for a new ERP system, it sounds like a problem for the IT department to resolve. Why should the guy loading the trucks in the warehouse really care? When replacing older systems, there are usually many improvement opportunities and software capabilities supporting business strategies. Look for them.

- **We need better information to manage the business.**
  - This is fine, but it is also nebulous and sounds as if we are fishing for a magical piece of information that will suddenly enlighten us all. From a business standpoint, what type of information is needed, how will it lead to better decisions, and most importantly, how will this help improve business performance or competitiveness?

- **We need an integrated system.**
  - Who doesn't? Said another way, what are the business problems today caused by lack of process and system integration, and how do these issues inhibit achieving business goals?

- **Corporate said we have to do it.**
  - The challenge here is how to turn this edict into something to rally the troops.

- **The ERP software will make your jobs easier.**
  - This statement casts the project in the wrong light. ERP is not about "easy" or "slick." It is about working smarter with better tools. When the goal is to improve productivity, it is best to state the business case in these terms or how a new system improves service to customers.

- **If we do not install a new ERP package, we will go out of business tomorrow.**
  - While a very compelling need for action, this is probably untrue. The business case must also be believable.

Project Objectives

During the project, the implementation team will face many tasks, issues, and decisions. The existence of clear objectives enables the team to set priorities to focus on what delivers the benefits.

If senior management does not set and communicate objectives, others in the company will. Different agendas and priorities will creep in to dilute the project focus, perhaps to the point of no one gets what they want.

Developing a list of objectives alone is not good enough. If not careful, it can become a pie-in-the-sky exercise where ERP becomes all things to all people, with no one truly committed to making it happen. The project manager must not allow this to occur because he or she will end up responsible for making all dreams come true!

Starting with the business case, the definition of objectives is a top-down and bottom-up approach in terms of the level of involvement. Though senior management establishes the agenda with the business case and initial objectives, gathering input from key functional managers serves as a reality check, fosters ownership, and adds more detail. At the highest-level, ERP objectives fall into these basic categories:

- Improve customer service
- Improve productivity
- Reduce process cycle times/lead-time
- Reduce cost
- Improve existing business strategies or enable new ones (intangibles)

When completed, ERP objectives should be:

- **Specific and quantifiable.** For example, the statement "improve productivity" does not tell us much. However, "cut order processing cycle time by 25%" or "improve on-time shipments by 20%" is something we can grasp.
- **Agreed upon** (by senior management).

- **Measured.**
- **Aggressive, yet achievable** (establish stretch goals, but not to the point of demoralizing those responsible for achieving them).
- **Documented and communicated** (enlisting stakeholders).
- **Owned by the right employees** (accountability to make them happen).

Benefits/Savings

At some point, a financial return on investment (ROI) calculation or some type of analysis of project costs and savings is usually necessary. Though there is plenty of independent benchmark data available regarding ERP benefits or savings, it cannot be emphasized enough to analyze your current environment and business practices to identify opportunities specific to your company. Basing benefits solely on the industry rags, as many often do, does not mean the company can achieve the same outcomes or that anyone believes the numbers.

Benefits are considered tangible or intangible. From an ROI standpoint, management is typically looking for the hard numbers. While there are usually cost savings associated with information technology support, system integration, and automation of business processes, many ERP systems cannot be justified by these alone from a financial standpoint.

This requires us to return to the original premise behind ERP: *ERP systems provide tools for better planning and utilization of resources.* This enables companies to deliver the right product or services to customers at the right time, at a lower cost. Hence the name: *Enterprise Resource Planning.* This is what separates ERP from other applications focused primarily on automating tasks.

If senior management does not understand the basic concepts of how these benefits are achieved with ERP, the company might be better off not pursing a new ERP system unless there are some very compelling intangible benefits. The following are areas to investigate when searching for savings that may impact the financials:

- Fewer shortages/missed sales (increased revenue)
- Reduction in inventory
- Reduction in inventory handling/carrying cost
- Reduction in product or material obsolescence

- Fewer equipment/machine changeovers (less expediting)
- Reduction in transportation/freight cost
- Reduction in unplanned overtime
- Reduction in material, labor, and overhead (negative) variances
- Less manual and exception-based processing and rework
- Reduction in outstanding accounts receivable
- Fewer customer invoice deductions/credits
- Reduced training cost for new employees or job changes
- Reduced IT hardware and software support cost
- Potential Headcount reductions

The Question of Headcount Reduction

The relationship between increased productivity and potential headcount reduction is this: ERP increases the "throughput" within the business—time to produce one unit of output is less using the same number of employees (in both administrative and operational areas).

Theoretically, management must then decide to take the cycle time (productivity) improvement or headcount reduction or reallocate resources to other important tasks. The answer depends on the competitive drivers within the business.

The relationship between increased productivity and headcount reduction is typically not one-for-one. This is worth noting since many organizations significantly overstate the savings from headcount reductions because they fail to recognize some of the subtleties.

For example, without question there will be at least some unforeseen software limitations that dampen benefits. Second, in order to take advantage of the new software capabilities or address business process related issues, *new* value added tasks are usually required. This means some productivity improvements are partially offset.

Third, productivity improvements many times cannot be easily isolated to specific jobs. Rather, much of the total time savings are *cumulative* across many departments, processes, and people. Yes, one can calculate a full-time equivalent (FTE) headcount reduction by adding the savings in each area. Nevertheless, in some cases the need for separation of duties (internal controls), and given the other tasks and skills-sets of those involved, may prohibit reallocation of their work in order to realize a headcount reduction.

Due to these factors, the headcount reductions achievable are usually less than the FTE calculation. For example, if the estimate is six FTEs, plan for three headcount reductions. Write off the remaining theoretical headcount reduction to productivity improvements.

In terms of IT department headcount, it is usually the same number of employees, using different tools, or doing different things. For example, when moving away from customized systems, you may need fewer software programmers, but this might create a new need for people that understand how to configure the software. This application analyst role is a different skill set, and usually the position ends up in the IT group. Of course, every situation is unique and there might be headcount reductions in IT if:

1.   The current systems are decentralized, supported by multiple IT groups, and the plan is to centralize support within one system.
2.   You currently have older systems with *many* people in IT writing software modifications and enhancements.
3.   The IT function is to be outsourced with the new system (which is an entirely different topic with its own set of advantages and disadvantages).

Without question, take the opportunity to assess IT department skills, determine resource needs with the new system, and evaluate the ability of those in IT to adapt to the new technologies. Also, keep in mind if there is little internal IT support today, there may be a need to increase headcount in IT.

Realizing Benefits

The ultimate goal of any project is to deliver on the benefits, whatever they may be. While managing the implementation costs and schedule are obviously important, if we fail to achieve the desired outcomes, a software installation is about the only thing accomplished.

First, it must be realized that project benefits do not begin immediately after the cutover to the new system. There is always an initial post go-live shakeout period. Usually at least one complete business cycle must process through the system until we can conclude most major go-live issues are behind us (ending with a smooth month-end financial close). Even on a well-executed project, it can take four to six months before users feel comfortable with the system and benefits start to kick in.

However, often there is no definition or a misguided understanding on who is responsible for achieving the benefits. While we would like to say *everyone* is responsible, in practice this approach does not always work. There must be some accountability for realizing each benefit.

In this regard, the first mistake is placing the responsibility on the wrong people, for example, the IT department, the project manager, the team, or the outside consultants. While all these groups certainly play important roles in benefits realization, none has direct control over the business processes affected by the new system.

Benefits realization starts with senior management. Each senior manager on the steering team should champion a benefits objective. The role is not to design or edict solutions but to provide oversight and address the barriers standing in the way of achieving benefits.

While every employee should feel accountable for project success, if middle management does not embrace the new system, benefits will probably not materialize. Middle managers are the process owners and set day-to-day direction within their areas. They must be heavily involved in defining the ERP benefits, eventually take ownership of the system and new processes, and make it happen. If they cannot do it, who will?

Measuring for Success

A common mistake is to assume that the mere existence of better software tools automatically results in a change in employee behaviors. The right performance measurements go a long way in encouraging the desired behaviors and reinforcing the new way of doing business and proper use of the system. Conversely, the wrong performance measures will continue to drive the same sub-optimization, with or without new software.

Also, keep in mind, informal measurements can destroy the effectiveness of any formal measurement system regardless of how many banners and slogans we hang in the workplace. This is about culture, and sometimes with ERP, cultural changes are necessary to be successful.

For example, if the production control manager gets a big pat on the back for constantly firefighting and expediting costly rush orders through the plant, how much emphasis will he place on correctly using the production planning capabilities of the new system? Probably not much. In this example, the manager is informally rewarded for *not* planning.

Furthermore, the adoption of new performance measurements is not a witch-hunt: It is identifying the reasons for less than desired performance and taking corrective actions. This means fixing the root causes of process deficiencies. Measurements are also a source for celebration (and perhaps monetary rewards) when goals are achieved and sustained.

Measurements should be quantifiable, with an understanding of the current level of performance, and with *goals* reflecting the desired level of performance. Measurement systems should address customer service levels, productivity, quality, and asset utilization. Measuring the integrity of the data in the system is also important.

For the purposes of ERP, it is usually not necessary (or advisable) to invent your own unique measurement systems. For example, measures for cycle time (productivity), first pass yield (doing it right the first time), inventory turns (utilization of assets), on-time shipments (customer service) and data accuracy, have been around for years.

The other problem with designing highly customized measurements is the tendency to exclude certain events or data under the premise the organization cannot control the root causes. The truth is there are very few business issues an organization cannot address or at least influence. If not careful, measurement systems can evolve into a game of fixing or fudging the numbers, taking the focus away from improving the business.

Some measurements (cycle time and first pass yield) can be applied to specific business processes. Again, processes are how the work is accomplished in any organization. Therefore, key business processes should be measured.

It is also important to recognize and manage the interrelationships between different types of measurements since most are not standalone. For example, if the goal of reducing product inventory drives inventory too low, customer service (on time shipments) will likely suffer.

There must be a balance to avoid having different management *hot buttons* driving the business every month. That is, one month the goal is to cut inventory and the next month it is to reduce product shortages (by building more inventory and working overtime to do it)! This is no way to run a business.

In order to realize all performance measure simultaneously you must fix the business processes and get the system working for you, not against you.

Other types of measurements should relate directly to the data that drives the entire system. If the foundational data in the system is not accurate, this affects many departments. The result is the system cannot be used effectively. Depending on the type of business, examples include order accuracy, inventory accuracy, and the accuracy of bills of materials. These measurements can be captured and reported through reoccurring system audits.

CHAPTER 9
CHOICE OF METHODOLOGY

Methodology Matters

Webster's dictionary defines methodology as "A body of methods, rules, and postulates employed by a discipline: a procedure or set of procedures." When it comes to ERP, many organizations commit to a particular implementation process with a limited understanding of what they committed to and how it may affect the project outcomes.

As mentioned, every ERP project entails discovery, design, construction, and testing, but how and when these occur might be different, depending on methodology. It is important to understand and select the right methods for the situation since it drives specific deliverables, the level of team participation, and the basic philosophies for managing the project. Many companies give vendors too much latitude in making methodology decisions or fail to ask the right questions in the first place.

There are no universal definitions for ERP methodologies, and there probably never will be. But in general, there are two ends of the spectrum: *traditional* or *rapid deployment* approaches. Each has its own set of advantages, disadvantages, and associated risks. The purpose of this chapter is to understand the decisions at hand and to provide guidance in choosing the right approach for the situation.

Rapid Deployment Approach

The concept of rapid deployment evolved to address the age-old problems: ERP takes too long and cost too much. As the name (or similar names) imply, it is geared toward getting there faster and cheaper.

In its purest sense, rapid deployment focuses heavily on several project management strategies. To a certain extent, these strategies can and should be used when managing any project. It is the degree and emphasis placed on them that sets rapid deployment apart from other methodologies.

Rapid deployment strategies include:

1. **Creating a sense of urgency by compressing the project timeline to expedite decision-making and issue resolution.** On many projects, valuable time is wasted in these areas.

2. **Relying heavily on pre-configured industry templates to drive the majority of the software set up.** The basic assumption behind the use of a template is that most companies within the same industry are similar in the way they operate or should operate. Theoretically, this supports the use of best practices and reduces the overall time to configure the software.

3. **Minimizing scope by cutting out tasks that get in the way of quickly constructing, testing, and going live with the software.** This typically means:
   a. Up-front project team training is scaled back (or omitted) and is mostly done informally when setting up the system.
   b. Up-front analysis of current processes and the design phase are reduced. Therefore, there may not be an *As-Is* or *To-Be* process deliverable as traditionally defined, since these activities become somewhat blurred with the actual construction of the system.
   c. There is a *zero tolerance* for software modifications. Even other types of custom programming can be highly discouraged (data conversion programs and interfaces).

4. **Utilizing a high concentration of consulting resources, but planned for a shorter period of time.** The premise is, the shorter project duration reduces overall consulting costs.

Rapid Deployment Risks

The potential drawbacks of rapid deployment are many. While the goal of any project is to complete it sooner, rapid deployment could become the old "slam-dunk" approach in disguise. If not careful, the outcome will be many issues and poor user acceptance of the system.

In addition, the cost savings with this approach is not a given. With rapid deployment there is a much greater chance that consultants will perform work that the client should perform. This incrementally increases consulting costs compared to other approaches.

Also, significantly reducing the scope of custom development can place unreasonable demands on users. For example, instead of automating many of the data conversions, users must enter a large amount of data into the system manually or hire temporary data entry clerks to do it. In addition, sometimes software modification can be easily justified from a business standpoint.

Furthermore, less involvement of the client project team is often an unfortunate by-product of rapid deployment. This can lead to less knowledge transfer, less input from users, and more employees resisting the change. This costs time and money one way or another.

Finally, with many consultants working on the project, the hope is the implementation is actually faster. If for some reason the rapid deployment techniques fail to compress the timeline, the project may cost a fortune.

Traditional Approach

The traditional methodology borrows concepts and practices from engineering, software development, and quality disciplines. The approach includes a well-defined set of project phases and deliverables. In simple terms, the idea is to "measure twice and cut once" with the goal of installing a quality system while minimizing costly rework.

While the traditional approach can include the use of industry templates for initial software set up and recognizes the value of prototyping, it places an equal emphasis on up-front activities such as team training, requirements definition, and a distinct design phase. These occur before the "official" software configuration (construction) phase begins. The concept is not to get too far along with building the system without a decent blueprint.

Finally, this methodology places a high priority on user acceptance of the system, thus assumes a deeper level of involvement of the project team and other stakeholders.

Traditional Approach Risks

The traditional method is not without risks. Some of the things that make it appealing can become detriments. There is the possibility of spending more time and money, yet still ending up with a lousy system. If not managed, potential problems include:

- Valuing *form* over *function*. Excessive formalities, deliverables, and documentation can bog down the project or only create the illusion it is on-track.
- The tendency to reinvent the solutions wheel, thinking "We are very different from other companies."
- "Paralysis through analysis" in a futile attempt to develop the perfect solution.
- Taking too much time to complete project tasks, make decisions, and resolve issues.
- Focusing so much on user acceptance of the system that the project takes on a life of its own.
- An excessive number of software modifications and other custom programs are permitted. Once Pandora's Box is opened, it is difficult to close.

The Best Approach?

The choice of methodology is as much a project management philosophy as it is a set of implementation steps. The preference of one over the other depends on the project objectives, the organization, and the tolerance for risk.

On large projects, those with complex requirements, or when a financial return on investment is the primary measure of success, think twice about using rapid deployment. These types of projects already imply more risk. In this situation, I would much rather face the potential for additional time and cost versus a greater chance of a system disaster at go-live.

In addition, the traditional approach provides for more opportunities to rethink business processes versus counting primarily on software to deliver the benefits.

When in doubt, lean toward the traditional approach. Remember that the steps and deliverables exist for a reason. Many within the ERP industry claim rapid deployment is *always* the right method because the traditional approach is too rigid. The truth is any methodology is only as rigid as you make it. My belief is that the many years of poor project management within the industry has unfairly given the traditional approach a bad name.

For example, if a project manager does not understand that the value of the "As-Is" analysis is more than drawing pretty flow charts or allows this phase to drag on forever, it is a waste of time! The same is true for formal up-front project team training when not properly managed. Also, defining business requirements early in the project does not imply that the requirements should not be allowed to change. Of course, requirements should (and will) evolve and change, but it is best to get some level of consensus on the business needs before we dive too deep into the software detail! In fact, not until the prototype, design, and construction phases are complete, will there be a solid understanding of the system requirements.

Rapid deployment or not, a good project manager contains scope, limits the amount of custom programming, sets aggressive schedules, uses implementation tools, does prototyping, and expedites issue resolution and decision-making. These can be accomplished without blurring or cutting out steps, or reducing project team involvement. You can deliver a quality system while only incurring the time and costs necessary to be successful.

Of course, this is not to say rapid deployment is never the right approach. For whatever reason, when the *overriding* business concern is to install the package as soon as possible, lean toward rapid deployment. There are legitimate business reasons for this, but most tend to be strategic in nature and less financial return on investment oriented, such as:

- The need to react quickly to customer edicts or regulatory requirements that are addressed with new software.
- A time-sensitive need to incorporate a recently acquired company into your existing ERP system.
- The "green field" start-up scenario (there are no existing business processes or systems).
- Smaller organizations where processes are mostly manual or there are only a few standalone systems to replace. In this case, the automation of tasks or just having one integrated system will be a major leap forward from an end-user perspective.
- System upgrades when the software functionality additions or changes in the new release are not significant.

CHAPTER 10
PROJECT SCOPE AND BOUNDARIES

What Do You Mean It Is Out-of-Scope?

The definition of project scope is one of the most important deliverables of the planning phase. Scope decisions drive all remaining plans, set user expectations regarding what they will and will not get within the software, and are an important tool to control the project.

A project scope that is ambiguous, fails to address all *dimensions* of scope, or lacks consensus or senior management sign-off, eventually results in schedule and budget overruns (scope creep) or last minute acts of desperation to get back on track (scope cuts).

In controlling the project, the project manager is similar to a referee in the football game and needs the ability to throw the penalty flag. Scope is that flag. It is always best when the rules of the game are clear to everyone involved before the ball is put into play.

It is critical for the entire project team, stakeholders in the functional area, and senior management to be involved in scope decisions. This can be a balancing act since the objective of any project manager is to limit scope to what is truly required to achieve objectives.

If the project becomes all things to all people, it will fail to meet anyone's expectations. Second, most projects have a propensity to get bigger (not smaller) with the passage of time. Therefore, if you do not get senior management sign-off on anything else, get it on the scope.

Taking Inventory

One of the first steps in this analysis is to develop a list of all business processes that *could be* within the project scope. As a starting point, the software consultants and vendor should have a generic list of processes for organizations within a similar industry or from other projects. If nothing else, begin by simply listing the major functions performed within each department.

When developing the inventory of current processes, go one level below the *major process* level. For example, in just about every organization, *procurement* is a major process, but this alone does not tell us much. When we break this process down to the next lower level, it starts to have meaning.

For example, within procurement, the sub-processes include: 1) Supplier Selection, 2) Supplier Master Set Up, 3) Requisition/Approval, 4) Purchase Order, 5) Receiving, 6) Invoice/Voucher Match, and 7) Payment. Later, when finalizing the business processes included in the project, any major variations to the sub-processes should be recognized.

Next, take an inventory of the existing systems. This is a list of all applications currently in place and important information for each such as the key user groups, major software functionalities, interfaces, and the technical platform (operating system, database, and hardware).

An excellent way to present the relationships between current systems is a high-level (context) diagram showing each system, key data interface pipelines, and the other information listed above. A picture is worth a thousand words.

When taking inventory, do not forget the one-off databases in the user area not supported by the IT department. These may include departmental stand-alone packages, custom developed databases, or important spreadsheets. Even though not supported by the IT group, treat these systems as relevant since some might be critical for daily operation of the business or for management reporting purposes.

Aligning Scope with Project Objectives

This may seem obvious enough, but in many cases, project objectives are aggressive yet management limits the scope to only what is necessary to get the software running on the computer. Perhaps worse, the project manager allows extraneous agendas to creep into the project scope.

The correct method in establishing scope is to understand that most organizations implement ERP to achieve some major business objectives (as defined by senior management). These objectives should drive the overall project footprint.

Of course, in order to implement ERP, some items should be inherently included in scope. For example, there are some current systems that must be replaced for the sake of systems integration alone. Beyond this consideration and the goals of senior management, any additional project scope is a negotiation with other key stakeholders. This process is always give and take, but some additional scope can sometimes solidify user buy-in and add benefits to the project for little additional time and cost.

The Ten Dimensions of Project Scope

The scope of any ERP project can be viewed through different lenses or dimensions. Each dimension is important in order to understand what the project truly entails. Scope dimensions include:

- Number of Sites
- Business Processes
- Software Modules
- Software Features and Functions
- Data Conversions
- Interfaces
- Number of Reports
- Software Modifications
- Number of End-Users
- Scope Boundaries

When developing scope, first address it from the standpoint of the entire project. The software rollout strategy (discussed in the next chapter) can affect the scope of each phase. A phased rollout usually takes on additional tasks to accommodate temporary or interim situations between phases. Therefore, one cannot get a complete picture of the project scope until the software rollout strategy is finalized.

The purpose of the discussion that follows is not to imply good or bad scoping decisions since this should be judged on an individual project basis. The purpose is to understand the dimensions, potential issues, and implications.

Level of Effort

Each dimension below should include a list of in-scope items, a brief and clear description of each, and the estimated *level of effort* (LOE). LOE is an educated guess (high, medium, or low) considering two things: 1) the shear amount of work involved or 2) the degree of complexity.

If nothing else, the estimated LOE is much better than identifying scope items with no understanding of what each major task might entail. Though somewhat subjective at this stage, the level of effort aids in developing a realistic schedule and budget. Within each scope item below, LOE considerations are addressed.

The Number of Sites

Many companies have more than one geographical location and/or have more than one *business unit*. A business unit is a division within a larger enterprise. In general, it has separate management structures, processes, employees, and separate financial reporting. For example, a business unit might encompass certain product lines or services offerings.

An implementation site is defined as a combination of a business unit and geographical location. If a single business unit has two geographical locations, it has two sites.

Alternatively, a single geographical location may house three separate business units in which case there are three sites. These are sometimes referred to as *logical* sites since each business unit may operate independently, but under the same roof. Often there are some shared centralized resources in this case.

Multi-site implementations increase the scope, level of effort and risk. First, there are more management teams to engage such as each business unit may have its own engineering or purchasing departments. Therefore, the project requires more coordination, communication, and consensus building among stakeholders. In addition, a certain amount of politics and cultural issues between sites is not unusual.

From an IT standpoint, more sites usually mean more legacy systems. This translates into more data conversion and interface programs to write.

All of these multi-site challenges can lead to project inefficiencies, slower decision-making, more software to set up, and more custom programming. A phased software rollout by site can mitigate some of the risks mentioned—but not all.

All scope items that follow should be identified by site.

Business Processes

The starting point for this analysis is the inventory of current processes. When determining the processes in-scope, there are several items to remember.

First, do not forget the *trickle down effect* from upstream processes. For example, if one of the goals is to significantly improve order entry productivity within the customer service department, this may not be possible unless the sales quotation activities are also within scope (this involves the sales department). Issues within the sales quotation process may routinely delay order entry or cause significant rework. We are now talking about internal supplier and internal customer relationships and the quality of information hand-offs. If the project fails to address the sources of bad or disjointed information, this may inhibit benefits realization.

In addition, a given business process *name* may actually represent more than one *physical* process. Again, take sales order entry as an example. Some orders are probably entered manually, some transmitted through EDI, and others via the web. Also, there could be international orders with a different twist. These processes might be performed by different departments or vary to the extent they should be treated as *separate* business processes.

While aware of these process variants, when not specifically acknowledged within the project scope, the schedule and budget can become grossly underestimated. From an implementation standpoint, there may be four or five order entry processes, each requiring unique analysis, software set up, testing, and training.

Degree of Business Process Redesign

The importance of estimating the level of effort was previously discussed, but business processes require special attention in this area. Each process included in the project begs the questions: How convoluted is the current process? Does it need significant improvements? Is there a goal to standardize the process across multiple sites? Is the plan to incorporate the current process into the new software or to implement an entirely new operating philosophy?

There is no right or wrong answer to the questions above. However, when significant process redesign is anticipated, there are more unknowns, issues, and decisions. This means more time is required to implement.

The Question of Standardization

As mentioned, a potential twist to the process redesign theme occurs when the scope includes the same business processes at multi-sites. In this case, at least a philosophical decision is necessary regarding the degree of standardization of the software at each site. There are two ends of the spectrum, and challenges and drawbacks are associated with each.

On one extreme is the notion that two or more sites will conform to a standard way of doing business and the software will be configured the same for all of these sites. The other end of the spectrum is allowing each site to "do its own thing" within the software. Of the two extremes, the latter represents more duplication of project effort, and higher IT support cost.

The philosophy of standardization is the right direction when certain operations and processes are truly similar. However, forcing standardization just for the sake of it (as so often occurs) is not a wise decision.

When the goal is to standardize, the philosophy should be to standardize *where possible*. In practice, different customers, products, and physical workflows may necessitate different operating procedures. This drives some unique site requirements and, therefore, some *localized* solutions in the software.

The standardization strategy boils down to identifying the system requirements common among all sites, while acknowledging those truly unique to each site. This is the *model company* approach discussed in the next chapter.

Software Modules

Determining the sites, business processes, and software modules within scope should be the first steps of any scope analysis. However, similar to business processes, it can be observed that some modules are inherently more difficult to implement than others.

As an example, for most organizations the purchasing module is usually easier to install than the sales order module, and the general accounting module is probably less difficult than warehouse management.

In addition, any module with the word "advanced" in front of it means more complexity, more time, and more money. This begs the question: Should advanced modules be included in the initial scope? Perhaps the functionality within the standard module offering in each area will suffice.

This decision must consider the sophistication level of the end-users. Initially, the standard module might be enough of a challenge as well as a major step forward in software capabilities (compared to the current system). It is true that sometimes users must learn how to walk before they learn how to run. In some cases, the advanced modules might be appropriate only after the users understand and successfully use the functionality within the standard module.

Software Features and Functions

When identifying the software scope, many only list the software modules included in the project. But each module has optional features and functions within it.

For example, a sales order module might have the capability to track customer rebates, but the functionality is not a prerequisite to install the system. Without this level of scope definition, understanding what a software module entails is somewhat unclear.

The alternative to scoping software features is to let each team figure it out later, wasting time and money in the process. They could spend days or weeks investigating and setting up certain software capabilities that management has no intentions of using.

If certain software functionality appears to add business value, but a lot of testing is required to make the final determination, then include it in scope. By doing so, at least it is covered from a schedule and budget standpoint.

Data Conversions

At the time of *cutover* to the new system, there must be a clean conversion of information from the old system to the new system in order to operate the business effectively.

Adding to the challenge is the fact that business data is dynamic in nature, constantly changing during the course of daily operations. Therefore, most data must be converted to the new system immediately before go-live, not weeks in advance. On the other hand, some data is very static in nature and can be loaded earlier. In this case, any change in the data is update in both systems manually by the users until cutover to the new system.

The data conversion scope includes the major system files to populate in the new software (target system), where the data is coming from (source system or documents), and how it is going to get there (manually entered or automated conversion program).

It is important to realize that some data must be loaded manually. First, there may not be an existing automated system from which to convert the information. Second, the functionality and data in the new system could be so radically different from the legacy system that an automated conversion may not be practical. The same is true if the data requires a significant amount of clean up.

Even when the above is not the situation, the best data conversion plan of all is to avoid automated conversions whenever possible. Again, writing more custom programs increases the timeline and costs. It may require less effort to load the data manually than to define, write, and test complex conversion programs.

One strategy that can help minimize the number of data conversion programs is to *bleed out* the data from the legacy system wherever it is appropriate to do so. For example, a week prior to the cutover, pay all open supplier invoices. At the same time, stop entering new supplier invoices into the old system. Enter the held invoices into the new system immediately after go-live, and later process the payment.

Also, convert history records only when necessary. For periodic history inquires, most users can live with retrieving the information from the old systems until the history builds up in the new system. Another approach is to save history records off-line in some simpler form for user access, such as a report or downloads out of the legacy system.

Beyond the need to write custom conversion programs, the level of effort with data conversions includes the amount of data clean up necessary, prior to converting it to the new system.

In most cases, data clean up is something only the users can address since they understand how the information is used in the business (much better than IT personnel or outside consultants).

There are two good ways to tackle the data clean up issue. First, clean it up directly in the legacy system prior to running the conversion programs. Second, extract the data from the old system into an *interim work file* that has a few custom applications to make the clean up task much easier. Clean the data up in the work file and then run an import program to load the data into the new system. It is usually not the best approach to load bad data into the new system and then try to clean it up prior to go-live.

The LOE is also affected by field (data item) formats and usages in the new system compared to the old system. For example, is the field size smaller in the new system? Is the field format alpha in the old system but numeric in the new system? Does it have the same business meaning? If these are not issues, the data item can be mapped directly into the new system. If not, the conversion program or the data clean up effort must address these inconsistencies, adding to the overall complexity of a conversion.

Finally, the availability of *data mapping templates* and out-of-the-box data load programs within the new package are very helpful. Nevertheless, a custom program to extract the information from the old system is usually necessary to get it into the import/load processes available within the new software.

Interfaces

The time and cost to write and test interfaces to legacy systems (or new third-party bolt-on applications) is almost always underestimated. Worse yet, interfaces to support the transition to the new system (as part of a phased rollout) are *temporary*, so the effort is pure throwaway once the older system is replaced. The bad news regarding *permanent interfaces* is the IT group must support them indefinitely and interfaces are usually not 100% reliable, functional, or real-time.

Permanent Interfaces

Replacing most of the existing systems with the new software usually results in fewer interfaces of any kind. This is not to say the new ERP package is destined to take over the entire company, but any decision to retain an existing system should be based on facts, not emotion.

The decision to keep or replace a system involves trade-offs between: 1) software capabilities, 2) the cost of purchasing and installing the additional modules necessary to replace it, and 3) the cost of retaining the current system, including the need to write interfaces and continue IT support.

When comparing the functionality within the current system to that of the new software, often there are many capabilities within the existing systems that are not utilized. Make sure the comparison includes only what is needed.

In addition, users tend to glorify the existing systems, but upon close examination of the business processes and how the system is used, it might not be so great.

Finally, even if a few software modifications are necessary to make the new software on par with the current system, often this is better than writing interfaces, supporting the old system, and dealing with multiple databases.

Temporary Interfaces

As mentioned, temporary interfaces are usually necessary during a phased implementation until the old system is replaced by the new system. When installing the new software all at once (Big Bang), there are no temporary interfaces. Therefore, temporary interface scope and the software rollout strategy are intertwined and should be addressed concurrently. This topic is covered in the next chapter and with tips to minimize the need for temporary interfaces.

Interface Complexities

Consider the following when estimating the level of effort required for interfaces:

- What may appear as a single interface between two systems could require many interfaces and some may go in both directions.
- Similar to data conversions, field usage and format issues between systems can result in cross-reference tables and complex interface programs.
- The required frequency of the interface (real-time or period batch updates) drives complexity. In most cases, the more real-time, the harder it gets.
- Software modifications (to either system) may be required to add fields to support the interface.
- Different technology platforms or limitations may prohibit the best possible interface design.
- Potential for additional software (middleware) and hardware purchases to support the interface.

Number of Reports

ERP packages include many standard (canned) reports. Some of these are usable in the current form, some must be modified, and others will not be used at all. In addition, there is always a need for new reports and these must be developed from scratch. The accumulation of all reporting needs can represent a significant software development effort.

The good news is that report development is usually not on the project's *critical path* since this programming normally runs concurrently with other activities. In fact, starting to develop reports prior to the first round of formal testing is ill advised. Until that time, no one can say for sure what is needed.

When scoping reports, the goal is to understand the total number of reports, not specifically list each one. Also, it is unlikely all existing reports must be duplicated in the new system for the follow reasons:

- Some standard reports delivered within the software will do the job.
- Many reports written over the years are no longer required (no one really uses them).
- Some of the existing reports are replaced by on-line inquires or user downloads available in the new system.
- Some existing reports should be combined into a new, single report.
- Most packages come with basic end-user *ad-hoc* reporting tools, allowing the functional analysts or the power users on the project team to write some simple reports after go-live.

Disposition Reports

First, create a complete list of all reports used from all systems to be replaced, the report distribution for each (who gets them), and the frequency of distribution (daily, weekly, monthly, or on-demand). Typically, there will be other reports that users prepare off-line in spreadsheets. Some of these reports may end up within scope.

Once current reports are listed, request the users to identify the ones they must have with the new system. Ask them to prioritize the remaining reports, and identify those no longer needed.

Also, do not accept new report requests from users at this time. Furthermore, attempting to determine if the many standard reports within the new software cover existing reporting needs is probably not worth the effort. The great majority will not, and if it is later determined some standard reports work fine, consider it a bonus. The only exception are reports that also update the database. In most cases, it is best to modify these reports, not create new ones.

Now that there is a list of report priorities, any report considered high or medium priority should be included in the project scope. Most low priority reports will probably not be required or can be written shortly after go-live. It is not an issue if priorities and even the total number of reports required are not 100% accurate. At this stage, it is a given that there are many assumptions regarding reporting needs with the new system.

Next, finalize the scope by estimating the number of *new* reports that will be required. As the project moves forward, the team will acquire more knowledge of the business needs and the new system, and new reports will be identified. Estimate the number of new (unknown) reports to be roughly 30% of the number of reports previously considered in-scope. If a few reports are not accounted for, these can be developed soon after cutover to the new system. There will always be more reports!

Software Modifications

Before we get started on this topic, a few industry terms need clarification. The first one is a *software configuration change* (or *setup change*). This refers to the parameters, switches, or options within the software that are selected by the system designer. These alter the software functionality to support a business need, without modifying the internals of the program code. These types of changes are not *hard coded* programming changes and, therefore, are not software modifications by definition. Every program has a finite set of configuration options, so the flexibility of any system is limited. This introduces the potential need for *software modifications*.

Modifications (mods) refer to hard coded changes to the standard programs, system tables, screens, or reports within the organization's unique *instance* of the software. Therefore, in order to implement a new software release from the vendor, usually modifications must be rewritten in the package, or are replaced by new functionality within the new release.

Many in the industry split hairs in the definition of software modification versus an *enhancement*, but it is important to understand the difference. The terms reflect how a customization is incorporated into the system.

A modification means *direct* changes to the software code. An enhancement implies the new functionality is built around (and yet integrated with) the system with *minimal changes* to the existing programs. One should try to do an enhancement, not a modification, whenever a software customization is necessary. Theoretically, enhancements are less invasive and, therefore, are easier to retrofit into new releases. However, many times modifications (not enhancements) are the best way to achieve the desired functionality. For the purposes of the discussion that follows, the two terms are synonymous unless otherwise noted.

Are Modifications Always a Bad Idea?

It is well known within the ERP industry that software modifications can substantially increase the project timeline and costs. In addition, modifications may negatively affect software quality and vendor support, and make future releases more difficult to incorporate. These are all good reasons not to make mods.

While acknowledging the potential downside, many organizations understandably refuse to allow major software limitations to dictate how they run their business. This is, perhaps, a more realistic view since software is intended to support the business, not the other way around. While software modifications must be discouraged, sometimes they make perfect business sense.

The belief that software modifications are universally "evil" came mainly from software industry. In fact, vendors can make you feel like a criminal when you sheepishly tell them you modified the software. This is because software vendors want their customers to upgrade their systems and do not want modifications or anything else to get in the way. The idea is to sell related software and plenty of consulting services with each "free" release or upgrade.

First, do not assume that the functionality within a future release replaces the need for a previous modification. Even if it is actually delivered, modifications might be necessary to make the new functionality work! Vendors use buzzwords to describe new functionality, but many times the capabilities fall well short of what is required.

While it is true that modifications increase the difficulty of software upgrades, in many cases this issue is over-blown. First, many companies never upgrade their software. Others may upgrade only once over the entire life of the system. In addition, there are tools available that make retrofitting modified programs much easier.

Finally, while most vendors initially refuse to address issues associated with programs that were modified by the client, most would assist if the issue is critical. Other vendors include support of modifications as part of their service offering.

So, what does all this mean for the ERP project manager? First, if you really can go *vanilla* (no mods), without crippling the business, absolutely do it. No one is encouraging modifications. On the other hand, no package is infinitely flexible and configurable. This *zero tolerance* for modifications philosophy is fine for those that do not have to live with the software limitations. The last chapter of this book addresses this dilemma with methods to control software modifications.

During the scoping and planning phase, the general rule is to make no allowances for modifications. This is true even if allowances were made in the project funding document. Remember, the scope document is used to manage the project, not establish high-end budgetary numbers. There are several other reasons not to include modifications in the project scope.

First, at this stage it is too early to determine if a suggested modification is necessary. This usually cannot be determined with certainty until the design phase.

Second, if there are allowances for modifications, you will probably make them— whether necessary or not! Again, we are not trying to encourage mods.

Finally, if the steering team later decides to proceed with a modification, they now own the decision to increase scope, expand the schedule, and incur additional costs. This is the way it should be, a business decision based on the facts.

The only exception for including modifications in the scope, is when during the software evaluation, some major functionality gaps were identified within the best-fit package, extra time was spent to verify the gaps, and the steering team agrees to assume the software must be modified. While never the best way to start a project, this situation does happen. However, even in this case the need for the modification should be tested and verified again during the prototype and design phases. Next, it must be business justified and approved by the steering team before any modification is actually performed.

Number of End-Users

Whether there are twenty end-users of the new system or a thousand, the same project work must be completed. Nevertheless, a large number of users can affect the time required for project communication and coordination, while increasing the number of decision makers.

A more tangible impact is the duration of end-user training and the amount of training resources required. These resources may include more trainers, more facilities, and more equipment. It is best to get a handle on this when planning to avoid last minute surprises later.

Boundaries (What Is Out-of-Scope?)

Anything that is not specifically included in the project scope is considered out-of-scope. However, even after in-scope items are identified some important stakeholders may make incorrect assumptions regarding what in-scope actually means.

For each dimension of scope, it is always best to document the key items out-of-scope. These are items some have suggested, implied, or wanted in-scope, but are not included. This step more clearly establishes and communicates the project boundaries.

As an example, the inventory manager might say to the ERP project manager: "Janice, I thought inventory reporting included the use of mobile apps." A typical response, when these boundaries were not defined ahead of time, might be: "Sorry, Joe, that is not what we meant, nor is it in the budget."

Even though a project manager may win a particular scope battle once the scope is defined, he may lose in the end when some key stakeholders no longer support the project. It is best to get them on board early with any major out-of-scope decisions or plan on additional scope from the very beginning.

CHAPTER 11
THE SOFTWARE ROLLOUT STRATEGY

Rollout Options

The strategy to rollout the software across the organization is the next important planning decision. The first strategy to consider is *Big Bang*. In its purest sense, this entails installing all software modules at once. As the name implies, there is a higher risk in this approach since there is a greater potential for business disruption at the time of cutover to the new system.

The other rollout option is some type of *phased* strategy. With this approach, the possibilities are many, but here are a few examples:

- Install a *mini phase one* (or parallel pilot) and next go Big Bang.
- Install all software modules, but only at one site or location at a time.
- Install a limited set of software modules across one or more sites, and then proceed with the remaining modules across the same sites.
- Install all software modules, but specific groups of customers or product lines are cutover one at a time.

The specifics of the organization, project, tolerance for risk, and the skills and experience of the project team should determine the strategy.

Do Not Wreck the Business

The most important rollout objective of all is to avoid bringing the company to a screeching halt at system go-live. ERP implementations have wrecked many organizations temporarily and some permanently. If the company cannot deliver products and services to customers because of system issues, this can have a long-term effect on the business.

Generally, it is my opinion that Big Bang is only appropriate when there is a single site, limited business complexity, only a few system interfaces, and a small number of users. In this case, if things go wrong, the chaos is more manageable.

The highest cutover risk of all is to combine Big Bang with rapid deployment in a large or complex company. Though this seems unthinkable, many have learned this lesson the hard way.

Typically, Big Bang is applicable to smaller companies, but small does not necessarily mean Big Bang. The business processes within some small companies can be more complex than larger ones. Needless to say, if you wreck the business because of a system implementation, senior management and customers will not care if the company is big or small!

Phasing: Pros and Cons

Summarized below are the advantages and disadvantages of phasing:

Pros:

1. **Reduced cutover risk.** Phasing minimizes the risk of systemic business disruption immediately after cutover to the new system for three reasons:

    ○ A phased project is less complex to manage; thus, fewer issues fall between the cracks.

    ○ The amount of business change occurring at once is limited. This provides time for the organization to absorb the change before moving on to the next phase.

    ○ It allows adjustments to the software based on lessons learned. This avoids exposing the entire organization to a major problem.

2. **Opportunity for some earlier success.** The longer a project goes without delivering any benefits, the more likely it will lose management focus and attention.

3. **Internal resources are not required all at once.** Phasing smoothes the demand for project team resources.

4. **Reduced consulting cost over the entire project (when leveraging the knowledge gained from earlier phases).** More on this later.

5. **Limited Financial Exposure.** Phasing limits exposure should there be some reason to cancel the project early on.

Cons:

1. **The entire project could take longer.** This is due to the serial nature of the phasing, the need to write temporary interfaces, some software rework, and retesting between phases. In addition, a project that takes longer has more exposure to changing organizational priorities and turnover of key project personnel.

2. **Delay Complete System Benefits.** Phasing implies the entire integrated ERP system is not fully functional until the last phase is completed.

3. **User inconveniences.** During a phased implementation it might be necessary for some users to work in more than one system (new ERP and legacy systems) until all modules are installed.

Phasing Considerations

The design of a particular software package can constrain or enhance certain phasing options. The software vendor can help you understand the design relationships between software modules, the sequencing constraints, and interface capabilities available out-of-the-box.

In addition, attributes of the business processes and data may drive certain phasing scenarios, or make others less attractive. For example, if customers order products on a single order but the products are shipped from multiple sites, this could affect rollout decisions. If not all sites associated with order fulfillment are on the new system, temporary interfaces may be necessary or, in some cases, may be not practical.

In practice, the phase planning process is similar to a simulation with alternative strategies identified and evaluated. In many ways, it is a balancing act, and typically, there is no optimal choice since most scenarios have at least some drawbacks.

Offset the Downside of Phasing

Phasing is the more commonly utilized strategy since it involves fewer risks. The downside is a potentially longer project timeline and usually higher implementation costs. The following are ways to offset these concerns:

1. Minimize Temporary Interfaces

Certain phasing strategies typically yield better results than others do in terms of reducing the level of effort associated with temporary interfaces.

First, replace as many legacy systems as possible within each project phase. This will enable the project to complete faster and can eliminate the need for some temporary interfaces or reduce the complexity. Of course, this is a trade-off between interfaces and module scope within each phase. However, the idea is to minimize temporary interfaces or select the ones that are less complex to write (considering the potential modules in each phase).

When selecting the preferred interfaces (based on the possible modules to include in each phase), the type of data and frequency of an interface may affect this decision. For example, some may require only a loose handshake between systems, such as batch updates twice a day involving only a few data items. This is much better than an interface that requires tight integration and real-time updates. Seek the path of least resistances when it comes to minimizing temporary interface work.

Finally, when considering temporary interfaces, always remember—they are temporary. Therefore, keep the design as simple as possible even if this means a few inconveniences for the users. In fact, when a particular temporary interface is not demanding, request the users to keep the data in-sync manually until the old system is replaced.

2. Ramp Up Software Knowledge

Another phasing strategy is to install less complex modules and/or locations in the first phase, while increasing the level of complexity with each subsequent phase. This reduces phase one risk and provides the opportunity for the team to gain software knowledge and implementation experience before moving on to more challenging phases.

This is more doable than many think. As examples, the basics such as system navigation, menus, screen layouts, and running reports are usually common across the entire system. Though the configuration options are different for each module, the tools and methods to set up the software tend to be the same for all modules. Even in this area, once you learn how to implement one module, the next one is much easier to learn.

Leveraging knowledge tends to work best for personnel that will stay with the project until it is completed. Even for project roles that have different employees participating depending on the phase, it is wise if these people can continue in a limited role in future phases to help transfer knowledge.

It is likely there are several rollout scenarios that enhance this approach since the sequencing of certain modules may better enable knowledge transfer due to commonality in software functions. For example, the plant maintenance and shop floor modules of most ERP packages both use work orders, parts lists, routings, and work centers. In another example, the purchasing and sales modules might have similar order entry processing or use common related modules such as inventory, pricing, and others.

3. Keep the Core Team Together

This topic is very much related to the previous discussion. First, turnover of any kind is disruptive and starts the learning curves all over again. The steering team should do everything possible to avoid unplanned turnover. This is particularly important on phased projects since these may take longer and this increases the potential for personnel changes. Thus, it is a good practice to provide some monetary or promotional incentives for the team members to stay with the company.

4. The Model Company Design

When phasing the same software modules across more than one site, a common mistake is to design and configure the system for only the first site and worry about the others later. In the end, this slows down a project.

By considering the requirements of all sites that will use the same software, the system is initially configured to reduce rework and duplication of effort to implement future phases. This includes not only the software settings, but also consideration to data conversions, interfaces, and the design of system-wide data structures.

This is accomplished with a *model company* approach where key areas of the software are designed and set-up to support the overall company structure and requirements common to all sites, while accounting for the unique needs of each site.

From a practical standpoint, the model company should address the foundational setup parameters and data that impact many areas of the software. These will be difficult to change after the first phase is installed without significant rework and retesting.

Another major advantage of this approach is that it provides the opportunity to discuss standard workflows and procedures across sites where appropriate.

Later as the software rollout progresses, the model is adjusted and requirements unique to each site are fully incorporated into the system just prior to the implementation of the site.

Building the model company requires the participation of key players from all sites in *joint design* sessions. These sessions can occur during the design and construction of the first phase. Furthermore, understanding how the system should be designed is the important thing, not necessarily performing all the set up in these sessions.

In any case, the result is the project team will better understand the implications of various design alternatives, thus avoiding bigger delays and costly rework later. The Model Company process usually includes:

- Setting foundational or system-wide parameters (constants) that enable certain software features and functions.
- Defining company-wide data structures such as financial companies, warehouses, chart of accounts, and other corporate level data.
- Considering the design of data conversion and interface programs to make them more reusable for each phase.
- Review of configuration options for commonly used programs.
- Defining the usages of fields in master tables such as items, suppliers, and customers (as these relate to each site), and reserving limited fields for site-specific needs.

5. Schedule Concurrent Phase

If project resources permit, running two phases somewhat concurrently will reduce the total elapsed time for the project. This allows an early start on a subsequent phase versus assuming all phases are serial in nature.

Select concurrent phases that are more or less unrelated in terms of teams, processes, and system integration. Also, start the next phase at a point when the previous one is well underway, such as after the design phase is complete.

Parallel Pilots

Parallel pilots are tools to help mitigate risks and increase learning. A pilot should be considered when developing the rollout strategy since it can be the first step. The most effective pilots are those run in parallel with the current system.

The term parallel pilot is often misunderstood because there are two different types. The first is a "limited" parallel pilot used only for the purposes of further testing (discussed in Chapter 18).

The second type of pilot is considered part of the software rollout. This is a "live" parallel pilot (real—not just a test). Planning a *live* pilot is similar to a limited pilot in terms of selecting the right location. However, unlike a limited pilot, the new software is live in the sense there is no intention of turning back to the old system. Although the purpose of the parallel is to have the capability to do just that, should the implementation go terribly wrong.

With a live pilot, parallel system processing runs for a short period. When all looks good, legacy system transactions are discontinued and users are now running only on the new system.

CHAPTER 12
ORGANIZING FOR SUCCESS

The project team cannot be finalized until after the project scope is defined. Similar to picking a ERP software package, when selecting the project team you have one chance to get it right. Poor project staffing decisions, not enough resources, undefined roles, and lack of accountability can doom a project before it gets started.

Figure 4 depicts the implementation teams and their reporting relationships. The project organization includes the Executive Steering Team, Project Management Team, Application Teams, Technical Support Team, Key Stakeholders (and power users), and the Subject Matter Experts.

The *Executive Steering Team* is ultimately responsible for the project. If the project goes down the tubes, they have no one to blame but themselves. Although there is an Executive Sponsor, everyone on the steering team must be a project champion. The role is to empower the project team, guide them, and do everything possible to help them succeed. Later in this chapter, the composition of the steering team is addressed.

The *Project Management Team* plans and controls the day-to-day aspects of the project. As mentioned before, a project manager from within the company should lead this team. In addition, a PM consultant should provide planning, implementation expertise, and coach the project manager, but also must step up and take charge when it becomes necessary.

Ideally, the PM consultant's heaviest involvement is during the planning phase. Thereafter, the goal is to reduce the consultant's involvement to project oversight only. In addition, the application team leaders and applications consultants must initiate leadership in their areas of responsibility. If they are waiting for the project manager to dictate their every move, you have the wrong people in these roles.

A manager from the IT department should function as the IT Project Manager within the project management team. The role is responsible for managing the technical teams and technology vendors, thus allowing the project manager to focus primarily on the business and application side of the project.

As mentioned before, IT personnel can add more then just technology support since they should be able to provide project management assistance and knowledge of the business and current applications.

If the company has no IT department or it presently only functions as a basic frontline hardware or PC support group, I recommend an independent technical consultant in this role to provide a higher level of expertise.

The *Application Teams* are organized around various areas of the ERP package, usually a specific software module. Each team consists of a team lead, application consultant (in-house or outside consultant), functional analyst, site representatives (on multi-site projects), and an IT support analyst (usually a developer/programmer). The team leads, functional analysts, and site representatives should come from the functional areas.

The Application Team is where the rubber meets the road. Their focus is on learning the system, designing business processes, configuring the software, resolve business and application issues, defining custom software needs, testing, writing work procedures, and end-user training.

The *Technical Support Team* provides a stable and reliable system environment and provides IT support in areas such as software installs, applying bug fixes, security, hardware, and other system administration functions. Some technical roles should be assigned directly to the project and they coordinate other IT support as necessary.

The *Key Stakeholder Team* consists of a select group of functional managers and end-users (i.e., power users) from the departments affected by the new system. They participate as required, but should be recognized within the project structure. Their involvement is critical and they have a responsibility to support the project.

At predetermined points, stakeholders provide input, feedback, and validation of the software design. In short, they act as a sounding board for the project team. Those stakeholders designated as *power users* have a deeper level on involvement. Periodically, power users perform hands-on tasks such as data clean up, testing, support end-user training, and the software rollout.

It is the responsibility of the project manager and each application team to determine the right time to get the power users and other members of the stakeholder team involved.

When selecting power users for the stakeholder team, include a few informal user "opinion leaders," not just those that are the obvious choices. Remember, you also want to hear from those that might be a tougher sell. As a bonus, if they are involved and get on-board, other employees that may not be overly enthusiastic about the project will take notice, and also get on-board.

The *Subject Matter Team* is a mixed bag of specialists called upon when needed. It includes internal or external resources engaged to provide specific knowledge or perform very specific tasks. Each role may vary considerably and is narrow in scope. For example, subject matter experts might include educators, compliance experts, or those providing specialized technical knowledge.

The Core Team

The project management, application, and technical teams are often referred to as the *core team*. The core team is the heart of the project and many team members should remain in their roles throughout the project duration. This should be the goal on phased software rollouts involving the same module(s) within each phase.

Otherwise, it is best if the application team leader and functional analyst are different people for each phase. In this case, each phase addresses different areas of the business and software modules, and those from the functional areas affected should fill these roles.

When selecting the core team members on a multi-site project, strive to have a mix of people from all company sites and locations. This bodes well from the standpoint of company-wide project acceptance. However, the overriding concern is to assign the most qualified people to the core team, no matter where they come from.

When core team members are from different company sites, their focus should not be site specific. They must take a corporate perspective to ensure the software works for all sites and locations.

Multi-Site Linkage

When the same software modules are to be used by more than one site, there are additional organizational considerations. As mentioned, the core team's loyalties should reside with all sites; but each site should have advocates to make sure the project addresses the needs of their site.

Therefore, the project organization should recognize site roles and responsibilities. *Site Representatives* define site-specific requirements, act as a conduit of communication between the core team and the site, and help coordinate software rollout activities when the site is to go-live with the system.

Failure to recognize these site roles can result in poor communication, inadequate definition of requirements, and more resistance to change. For example, employees at remote locations might complain that, "decisions are being made by people at headquarters and others that do not understand our needs."

Within the organization chart, one can show site representatives as a separate team or additional team members within the core team when the site is to be implemented. In either case, they must work closely with the core team.

From each site, it is best to have one representative work with the project management team and other reps with the appropriate application team(s). When systems residing at various locations are affected by the project, a site rep from the associated IT group should work with the technical team.

Site representatives are usually not fulltime on the project. They transition in and out of the project at the appropriate times. The involvement of each site is the heaviest at three different stages of the project.

First, site reps participate in the beginning when developing the project plan. They are also involved in defining needs and setting up the model company, previous discussed. After the initial planning and model company set up, usually only the first site(s) scheduled for rollout continues to work day-to-day with the core team until the site is up and running on the system. Once the first phase is complete, the next site gets involved to help prepare for their software rollout.

Do Not Over-Structure

Like any project, we need the right skills to get the job done, but on many projects, there are too many teams and layers of management. Similar to too many vendors, more does not necessarily mean merrier. The goal should be to "right size" the project organization.

When a project is over-structured, there are many unnecessary hand-offs, redundancies, poor communication, and a lack of accountability. For example, in addition to the project manager, a separate project director is usually unnecessary except on extremely large projects. On most projects, the project management team, the executive sponsor, and steering team can steer the ship. The same is true for a separate group to handle change management.

In addition, a separate business analyst or best practice experts may cloud responsibilities with the application consultant and functional analyst. Again, this does not mean separate people should never fill these responsibilities, but before adding more people or teams, think carefully about why they are necessary.

Resource Commitments

When planning project resources, usually the first issue is senior management's initial shock of the time required of team members. This is where education on ERP, lessons learned, and understanding what the team will be doing is helpful for them to grasp the scope of work involved.

The amount of time the core team must dedicate to the project each week will vary depending on the phase and their role. Nevertheless, for the project manager, application team lead, functional analyst, and the "in-house" application consultant (if you have one), anything less than four days per week on average could get the project in trouble. If these roles spend less than three days per week on average, outside consultants would be running the project (something that should be avoided). The only exception to the four-day per week rule might be the technical support team and the IT support analyst. As mentioned, key stakeholders, subject matter experts, and site representatives are involved on an as-needed basis.

Again, these are only averages. At times, core team members might work on the project one or two days a week; at other times, working through the weekend might be necessary. ERP projects are simply too fluid to pin down specific time commits for each week for internal roles. However, getting senior management commitment for the average number of hours per team member per week is important.

The "B" Team

Once time commitments for each role are settled, assign the right employees to the project. As previously mentioned, your best people are not hard to find, they are your "doers" and "go to people". Every organization has them. If not, the organization could not stay in business or get much of anything accomplished.

In addition, package software has been around for a long time. Look for a few employees that were involved with a major software implementation at some point in their career. Take advantage of their experience if these employees want to participate.

When attempting to staff the project with the most qualified employees, count on resistance from the management ranks (especially from their immediate supervisor or manager). Getting the right employees involved is always a struggle, but it is the first test of senior management commitment. It also sends the message to the rest of the organization of how serious management is about the project.

Some in management will insist they cannot afford to give up their best employees for the project. Of course, these employees are always busy working on something because they are the best. The easy choice is to assign employees that just happen to be available, but this is a known recipe for failure.

Of course, no one can argue with the fact we must keep the business running and, no doubt, the project will squeeze internal resources. The good news is I have yet to see any business stop running successfully because of good employees spending a lot of time on an ERP project.

The staffing question for senior management is: Can you afford *not* to assign the best employees? Remember, the idea is to run your business on the software, the thing that enables delivery of product or services to customers and accurate financial reporting. The stakes could not get much higher.

In addition, ERP costs money, and there are system benefit expectations out there, perhaps even some unrealistic ones. The careers of a few executives are probably riding on outcomes. At the very least, no CEO wants to be in the unenviable position of explaining away an expensive ERP disaster to shareholders or the big wigs at corporate.

The business reasons for allocating the right employees are not just about the threat of failure. There are opportunities, savings, and other benefits of taking project staffing seriously. These include reducing consulting costs and positioning the organization to support the software and users after the system is in production. Otherwise, the company may not leverage the investment or become permanently dependent on expensive outside consultants.

Also, by the time the project is over, those directly involved acquire new skills and learn more about the organization and software than they ever dreamed possible. These are the future leaders of the company.

Finally, now is the time to develop the back-up personnel for key employees—something management has always envisioned. Transitioning in employees as back-ups to allow the right employees to participate is an excellent way for this to occur. These are your future go-to people.

At times, companies make some poor staffing decisions relating to the use of temporary employees on the core team. Hiring temporaries to fill application team roles such as Team Lead or Functional Analysts is not wise. Hired just for the project, these people are not familiar with the company and have no creditability with peers. Also, once the project is over, they walk away with all the software knowledge.

As mentioned, hiring a temporary Project Manager (when an internal person cannot be found) is different. This individual should bring both project management and software specific expertise, but does not require the existing business knowledge expected from a team lead or functional analyst.

Finally, ERP is a challenge and all team members must be up for the task. When presented with a choice, I would much rather have good employees that *want* to participate versus those that might be the best overall candidates but do not want to be involved. The sheer willingness to learn and do what it takes to be successful goes a long way in overcoming some skill set limitations.

Not Enough Internal Resources?

OK, so what if there are not enough bodies to backfill for your best employees to participate on the project? Often if you look hard enough, get creative, and stop doing things within the company that add no value; plenty of back-ups will magically surface.

If necessary, hire temporary employees to back-up the back-ups. This is much better (and cheaper) than consultants performing most of the project work while your best employees sit on the sidelines.

As discussed in Chapter 3, if there really are not enough resources or skills to staff the project, the answer is simple: Hire permanent employees with the skills. Again, organizations do this all the time for tasks much less important than ERP. Do not immediately assume new employees for ERP cannot be cost-justified.

Plan the Resource Transition

Once people are assigned to the team, we must make sure they will be available when needed and in the number of hours promised by senior management. It is important that the project manager, the employee, and the employee's immediate supervisor agree on how the transition is to take place and when. Therefore, meet with each team member and their supervisor to develop this transition plan.

The plan should identify all current responsibilities, the specific tasks to shift to other employees, and a date each will occur in order to ramp up the individual's availability. When all parties agree on the plan, it makes the transition much easier for everyone.

Project Team Accountability

On projects like ERP, when there is a lack of accountability, there is lack of success. Everyone on the project has roles and responsibilities to fulfill. We must clearly communicate what is expected, monitor results, and make necessary corrections.

Having said that, in the beginning, some team members will be uncomfortable in their project role since implementing software is not something they normally do. However, the right employees will adapt, grow, and become more confident every day. In fact, some will want to continue working with the system in a support role when the project is over.

One job for the project manager is to plan for additional help for certain team members that may need it. This means additional training or consulting support. Granted, there is a fine line between supporting someone and carrying them, but we cannot let team members fail when they are trying to succeed.

We also must create a project environment where team members ask for help when they truly need it. No project manager wants last minute surprises because someone was simply afraid to ask for help.

At the same time, a project manager must be a realist. When it is acknowledged that a team member is not going to cut it, find a replacement as soon as possible and move the employee back into their normal job. As a contingency, plan for this (and for the risk of some leaving the company) by thinking about who the back-ups might be and get them more involved from the start. This is also a concern for outside consultants. When it appears a poor choice was made in hiring a consultant, try to recognize this early and replace them as quickly as possible.

The Right Executive Sponsor and Steering Team

Ideally, the executive sponsor should be an executive or senior manager within the company that has direct authority over *all* business processes that will eventually use the new system. The sponsor should represent the highest level of "process ownership" from a project scope standpoint. This enables him or her to make things happen and not always have to convince others on the steering team, or within middle management, of the need to take some important action.

Of course, it is always best if the CEO is the executive sponsor. After all, the CEO is ultimately responsible for everything and has the ultimate authority. However, in larger corporations the CEO as the executive sponsor is not always practical.

At a minimum, the executive sponsor should be a senior manager (V.P. or Director) who owns the *great majority* of the business processes affected by the new system. This means the executive sponsor does not have to be the CEO since it depends upon the project scope.

In this case, I value the process ownership criterion over other senior managers at the same level that might have more skills to perform the sponsorship role. In the end, if the senior manager responsible for the area cannot (or will not) make the new system successful, it is wishful thinking to assume some other senior manager at the same level can be successful. Finally, in this example when in fact there are major concerns about the proposed sponsor, definitely go after the CEO for the role.

At the same time, recognize that the executive sponsor (no matter who it is) will not be superhuman. The individual may own the businesses processes, support the project, and be good at running the operation, but not so good at sponsoring an ERP project. However, this is one reason there are others on the steering team to offer their business knowledge, guidance, and influence. Finally, a strong project manager can offset some of the weaknesses at the steering team level.

The other members of the steering team should include the rest of the senior management staff, particularly those responsible for the other functional areas that are directly or indirectly impacted by the system. This should also include the head of the IT department.

Those that should be periodically involved on the steering team include the partner or practice leader of the consulting firm engaged, and the highest level manager within the location to be implemented (in a phased software rollout). For example, the plant manager at a manufacturing site.

Sponsor and Steering Team Responsibilities

The executive sponsor chairs the steering team, and the project manager facilitates the meetings. The role of the sponsor is to work closely with the project management team, and other senior managers to ensure success. The sponsor from time to time must also engage the rest of the organization and the project team, as well as get into the project details when necessary. The executive sponsor and steering team responsibilities are to:

1. Ensure that executives are educated, are on-board, and understand their roles.
2. Own the project business case, objectives, and drivers for the change.
3. Implement new measurements to reinforce the desired behavior and process changes.
4. Approve and contain the project scope.
5. Assign the right employees to the project team.
6. Free-up the required time for those on the team to participate.
7. Expect (not just hope) that the internal team and IT support eventually become software experts.
8. Hire employees with the right skills and knowledge when necessary.
9. Hold functional managers, the project manager, the project team, the IT staff, and consultants accountable for results.
10. Require (not just sell) the cooperation of employees at all levels of the organization.
11. Participate in the project kick-off meeting and on-going project communication and to show visible support for the project when the opportunity arises.
12. Make necessary changes in operating paradigms and business practices to take advantage of the software.
13. Limit software modifications through business justification or changing business processes.

14. Remove the people barriers and naysayers that stand in the way.
15. Tackle project related business issues and decisions in a timely fashion.
16. Take end-user training seriously and require employees attend.

The Project Manager—More Than Just a Figurehead

As mentioned previously, the project manager should come from a functional department within the company so the project is not viewed as externally driven or just another technology initiative.

Unfortunately, in many cases consultants or the IT department are left performing the work while the project manager becomes virtually non-existent. This is one of many reasons the selection of the project manager is the most critical staffing decision.

The position requires a solid manager, with the right foundational skills and capable of developing new ones. The individual must be from a high enough level within the organization in order to work with most other managers on at least a peer-to-peer basis.

The PM must work with individuals (outside the team structure) at every level of the organization and across departmental boundaries, thus should have the skills to navigate the department silos. Still, the PM will probably need some help from the steering team in this area.

In order for the project to be viewed as a company priority (not just the agenda of the project manager) it is always best to move the project manager out of their current reporting relationship and report directly to the CEO for the duration of the project. This helps elevate the importance of the initiative above any department, and creates "power through association" by raising the project manager's status within the company.

Project Manager Attributes

No matter who fills the PM position, it is a different role from that of the project management consultant. If for no other reason, you want someone (internally or an independent contractor) working solely on behalf of the organization when dealing with management and all vendors.

Like other key areas of project staffing, the requirements of the job should take precedence over who happens to available. So what are the skills we should look for in an internal project management candidate? While specific responsibilities clarify the duties of the PM, understanding these alone may not result in selecting the right person. An ERP project manager has certain traits and attributes that are the best predictors of success.

Below is a list of attributes any in-house candidate must have to succeed in the PM role, followed by desirable (though not required) traits and, finally, the pluses (traits that would be additional assets).

If unable to find the ideal internal candidate, remember that beyond the important attributes, a project manager coming from within the organization is an advantage, and therefore does not have to possess all of the other skills to be successful. Some of the weaknesses in the desirable traits can be addressed through training, and identifying where the executive sponsor, IT manager, and project management consultant can play larger support roles to help fill the gaps.

When no acceptable internal candidate can be found, look externally for an independent or a new employee to fill the PM position. When forced to go outside, we are looking for a slightly different and higher-level skills in certain areas. Ideally, we want all the skill-sets (including the "pluses") as described below (except those that obviously relate to only an internal person). An outside PM of high caliber can offset some of the drawbacks associated with not having an existing employee running the project.

PM Attributes: Must-Haves:

1. Demonstrated leadership abilities as a key functional manager in the organization.
2. Credible at all levels of the organization.
3. Experience managing cross-functional teams.
4. A track record of successful improvement projects.
5. A good grasp of the business, people, and culture.
6. Ability to communicate and present information effectively.
7. Capable of learning the software implementation steps and pitfalls.
8. A planner, not one who only reacts.
9. A "people person" but can take a "tough love" approach when necessary to uphold accountability.
10. Comfortable with using technology.

Desirable Attributes:

1. Previous experience with implementation of any software package.
2. Ability to get into the project details when necessary.
3. Able to view the organization as a set of business processes (not departments).
4. Resilient and can adapt to changing situations.
5. Able to grasp new concepts easily (a quick learner).

Major Pluses:

Listed in sequence from most important to least important:

1. Project management experience with the ERP package selected.
2. Application consulting experience with the ERP package selected.
3. Strong business analysis skills.
4. Project management certification.

Project Manager Responsibilities

The project manager leads and facilitates the initiative. In the end, the role is responsible for the following items (though supported by the steering team and others within the project management team to the extent required):

- Plan and control the project scope.
- Develop implementation strategies.
- Develop and maintain the project master schedule and budget.
- Communicate team responsibilities, assignments, and deliverables.
- Manage outside consultants and vendors.
- Develop and execute the project communication plan.
- Develop and execute change management strategies.
- Manage overall software quality and user acceptance.
- Identify and manage project related issues.
- Identify and manage risks to the project.
- Work with department managers to implement process improvements.
- Track and report project status.
- Maintain the list of software development tasks and responsibilities.
- Facilitate timely decision-making and issue resolution.
- Plan and conduct Executive Steering Team meetings.
- Elevate major issues and project barriers to the Steering Team.

Application Team Leader Attributes

The Application Team Leader should be a functional manager from the user area. The application team leader should be the direct *process owner* over most of the business area affected by a specific software module. For example, it is best if the team leader for the accounts receivable module is the accounts receivable manager.

A big reason for this approach is that if you own a process, you are in the best position to change it! In addition, as a manager, the person should bring at least general management skills to the project. Finally, a manager of the area can more easily get more employees involved when necessary.

The second choice for the job is an individual with strong business analysis skills, creditability, and has influence with the process owners and employees in the area. For someone without these traits and who is not the process owner, it can be very difficult for that person to be successful as the application team leader.

Functional managers as application team leaders may be of concern to some because middle management and supervisors can erect the biggest barriers to positive change. Admittedly, care must be taken to set the right expectations for the role. However, whenever middle managers refuse to support important company objectives, the first question is: Why are they still managers?

Furthermore, if a process owner does not support the initiative as a member of the project team, they certainly will not support it when on the outside looking in. From a project manager's perspective, there is an old saying: "I would rather have you throwing rocks *inside* my tent than throwing rocks *at* my tent". If a lack of support occurs within the project tent, the project manager is in a better position to address the issue. In addition, whether a process owner fully supports a business change or not, he or she is ultimately responsible for making it work within the business. That is, no one likes to throw rocks at themselves!

It is a common mistake to have the application consultant also serve as the application team lead. The team lead role is a "mini" project manager covering a specific area, not a software specialist. An outside consultant certainly directs and guides many team activities relating to the software but has no authority to manage employees or to make changes within the organization.

In fact, most application consultants welcome not having complete responsibility for team leadership. This allows them to focus mostly on what they do best… software. This leaves many planning and coordination activities, and most business issues and decisions, in the hands of the organization (where it truly belongs).

Application Team Leader Responsibilities

Consistent with project directions and with the application consultant as a coach, the role of the application leader is to manage and coordinate team activities, work with other affected process owners, and provide input into the software design. Responsibilities include:

- Learning the key software capabilities and how they related to managing the business.
- Organizing team activities to support the implementation process.
- Maintaining the team's task list, issue list, and meeting schedule.
- Ensuring that tasks and deliverables are completed on-time and in a quality fashion.
- Providing input into the design of business solutions (consistent with project objectives).
- Participating in project communications.
- Helping remove barriers to the team success.
- Ensuring all key stakeholders are involved.
- Assigning action items and managing timely resolution of issues and decisions.
- Elevating issues to the project manager when necessary.
- Reporting team status to the project manager.
- Working with the team to develop end-user training plans.
- Supporting end-user training.

Functional Analyst Attributes

Each application team has a Functional Analyst. The analyst is a knowledge-able user from the functional area of the business. This person works closely with managers, other users, and the application consultant to design and imple-ment the software within the business operations.

When using the services of an outside application consultant, the functional analyst must acquire a deeper knowledge of the system to provide end-user support after the consultant leaves. The functional analyst should have the following attributes:

- Has creditability with management and peers.
- Possesses a detail understanding of the current business processes.
- Is a "power user" with the current systems.
- Is a change agent (wants to make improvements).
- Has the ability to grasp new concepts and tools fairly quickly.
- Takes a hands-on approach (is not afraid to touch the keyboard).
- Is a problem-solver.

Functional Analyst Responsibilities

Many view the functional analyst as someone "representing" the user community on the project. However, there should be more to it than just providing feedback and rendering opinions since it also requires a lot of hands-on work.

Under the direction of the application team leader and with the application consultant as a coach, responsibilities include:

- Defining current processes and how existing systems are used.
- Getting educated on industry accepted business practices.
- Identifying improvement opportunities.
- Defining software requirements.
- Understanding the software capabilities, transaction processing, and becoming knowledgeable about software configuration settings.
- Identifying and documenting software gaps, project issues and decisions. Driving resolution from a business standpoint.
- Fostering the involvement of other stakeholders and end-users outside the project team.
- Working with stakeholders and the application consultant to design the new business processes.
- Working with the application consultant to configure the software. Document the configuration settings.
- Assisting with requirement definition for custom programs (data conversions, interfaces, software mods, and reports).
- Training "power users" so they can assist with the project.
- Participating in project communication.
- Providing input to project status updates.

- Developing and executing software test cases and scenarios.
- Documenting new work procedures (standard operating procedures/desk level instructions).
- Coordinating user acceptance testing with key end-users.
- Developing end-user training materials.
- Training end-users (with the support of the power users, application consultant, and team lead).
- Performing system cutover activities as assigned.
- Supporting the end-users immediately after system go-live.

"In-House" Application Consultant Attributes

This book previously covered selecting outside application consultants and the alternatives of hiring new employees for this role or developing the role internally. Chapter 16 addresses strategies for the team to acquire more software knowledge and even "grow your own" application consultants. However, sometimes a hybrid approach can make sense (e.g., using an outside firm for some modules and hiring or developing your own consultants for other modules).

Note that when developing an in-house application consultant or when hiring a new employee for the position, this person does not displace the role of the functional analyst.

Application Consultant Responsibilities

Of course, when developing an application consultant, support will be necessary from a consulting firm until the individual gets up to speed and can function effectively on their own. The key is how quickly the outside application consultant can back off into an "as needed" support role. Depending on the quality of the person being developed, in many cases the transition can happen quickly. For others it will take longer, and for a few, the level of expertise expected might never materialize. In any case, with this approach you should require less outside consulting support during the project and over the life of the software within the organization.

Therefore, the responsibilities listed below are joint responsibilities of the person being developed into the application expert and the outside consultant, until the outside consultant transitions into an as required role:

- Communicating the implementation steps for the assigned module.

- Analyzing current business processes and issues, and recommending improvements.
- Facilitating the definition of software requirements.
- Providing insight into the "best practice" application of the software.
- Transferring software knowledge to the rest of the team.
- Coaching the team in completing their assigned deliverables.
- Coaching the team in configuring the software.
- Identifying and documenting software gaps (capabilities vs. requirements).
- Leading definition of custom program specifications (data conversions, interfaces, etc).
- Coaching the team in developing software test plans.
- Assisting in resolution of business issues and decisions.
- Trouble shooting software bugs.
- Providing input to team status report and the issue list.
- Coaching the team in developing end-user training plans and support the training.
- Leading the development of detail cutover plans.
- Backing-up those providing the first-line of end-user support after go-live.

IT Support Analysts

Each application team should have the support of an IT analyst. An application developer from the IT department (not a technical person or outside developer) best fills this role. The position provides knowledge of current system capabilities and data, writes custom software programs, and is a conduit into the IT group to support other needs of the application team.

The IT analyst, unlike most roles, can be part of more than one application team. Their involvement on any one team is usually not full-time until the spike in development work begins, typically halfway through the project. At that time, additional software development support may be required.

If internal developers do not exist, someone from within your IT department should participate on each application team until custom development needs are defined. At that time, this individual can help manage the contract programmers when development work begins.

One question is: What types of vendors are best for custom software development—the primary consulting firm working on the project, the software vendor, or third party contractors? I recommend the use of third party contract programming resources, whenever possible, for several reasons.

First, usually one can acquire highly qualified contract developers for a fraction of the cost (when compared to the primary consulting firm or the software vendor). In addition, there is little risk using third-party contractors when they have a clear and approved programming specification prepared in advance, and someone on the application team is readily available to answer their questions.

Company vs. Consultant Responsibilities

While working together as a team is imperative, it is in everyone's best interest to understand the responsibilities of the organization versus those of the outside consultants. While one can argue that the company and the consultants have equal responsibilities, there are in fact different expectations when completing major tasks.

Addressing responsibilities for each project deliverable takes the mystery out of who is doing what to help avoid unnecessary schedule slippage and budget overruns. This analysis also aids in determining the type and amount of outside resources required. Finally, assignment of responsibility also reflects the timing of when consultants will transition off the project.

The goal of this analysis is not to split hairs, but to define who has *primary responsibility* for each deliverable. When one party does not have primary, they are expected to play an active *support* role.

In short, the party with primary responsible drives and facilitates the completion of the deliverable and may perform most of the "hands-on" work to complete the task. This *does not* always imply the primary party has to be the expert on the task. The support role is expected to do whatever is necessary to ensure the primary party is successful.

Every project is different, but considering the insights discussed in this book, it is likely most organizations can take primarily responsible for more than they initially believe. If the team stumbles at times because of taking on more than they can handle, the job of the consultants is to back them up. In fact, when serious about developing "in-house" application experts, this is the philosophy.

Figure 3 in Chapter 7 depicts the key project deliverables within each phase. This is used to assign the primary responsibility for completing each deliverable and who plays the support role.

Figure 8 – Project Organization

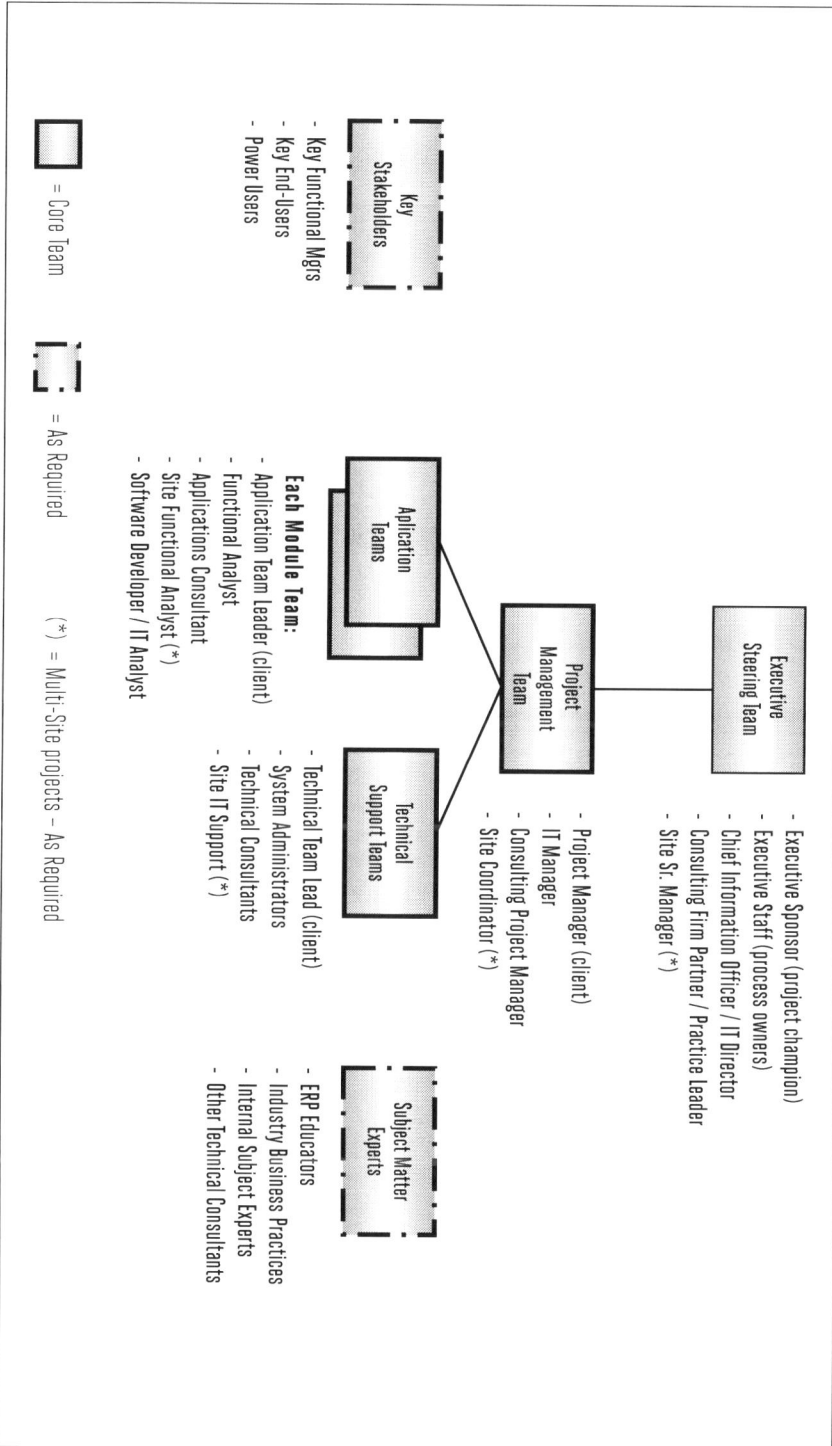

Executive Steering Team
- Executive Sponsor (project champion)
- Executive Staff (process owners)
- Chief Information Officer / IT Director
- Consulting Firm Partner / Practice Leader
- Site Sr. Manager (*)

Project Management Team
- Project Manager (client)
- IT Manager
- Consulting Project Manager
- Site Coordinator (*)

Application Teams

Each Module Team:
- Application Team Leader (client)
- Functional Analyst
- Applications Consultant
- Site Functional Analyst (*)
- Software Developer / IT Analyst

Technical Support Teams
- Technical Team Lead (client)
- System Administrators
- Technical Consultants
- Site IT Support (*)

Key Stakeholders
- Key Functional Mgrs
- Key End-Users
- Power Users

Subject Matter Experts
- ERP Educators
- Industry Business Practices
- Internal Subject Experts
- Other Technical Consultants

☐ = Core Team   ⬚ = As Required   (*) = Multi-Site projects – As Required

179

CHAPTER 13
THE REAL SCHEDULE AND BUDGET

In many cases, the ERP schedule and budget are not worth the paper on which they are printed. Many project managers shoot themselves in the foot by prematurely committing to a plan that only sets the wrong expectations. Others throw darts to come up with a go-live date, or falsely assume the quote from the consultants is the gospel.

The problem is once unrealistic expectations are cast, they are not going away. Could this be why so many ERP projects fail? In fact, some argue many projects are not really failures, just a failure to manage expectations!

For most senior managers, the project schedule and budget are usually their biggest concerns. Providing the best possible answers is part of a project manager's survival guide, but there are also other implications.

Unrealistic time and cost commitments drive poor project decisions. Many project managers cut corners in an attempt to catch up to a plan that was never agreed upon or feasible from the start. On the other hand, most teams will rally around a schedule if they had input in developing it and believe in the plan.

Three Stages of Estimating

Estimating the project schedule and budget usually occurs in three stages: 1) Developing early estimates before a decision to proceed with an ERP software evaluation, 2) When securing project funding after evaluating software and consultants, and 3) Formulating a baseline schedule and budget during the planning phase which will be used to manage the project. Each stage of estimating goes deeper into the details since there is more information available as the project progresses.

As with all estimates, always include your list of assumptions, constraints, and risks. If the project takes too long and cost too much, and this list is not documented, management will not recall all the things that drove the previous estimates.

Early Estimates

The project manager should anticipate that senior management would want at least high-level estimates before making any type of financial commitment to the project. There are many ways to develop estimates, but it will never be a science. Below are my general guidelines for a project timeline based on company revenue or the number of users:

- Smaller companies (under $100 million, or under 125 users) = 9 to 16 months
- Medium companies ($100 million to $250 million, or 125 to 250 users) = 16 to 24 months
- Larger companies ($250 million to $1 billion, or 250 to 600 users) = 24 to 36 months
- Very large companies (over $1 billion, over 600 users) = 36 to 48 months

Many factors will affect the actual timeline, but for the purpose of ballpark estimates, the four major assumption areas driving the ranges above are: 1) Project team time commitments (full or part-time), 2) The number of modules to install, 3) The software rollout strategy (Big Bang or phased), and 4) The amount of software modifications expected.

A quick, yet fairly accurate way to estimate total project cost is to back into the numbers based on the cost of the ERP software. For example, the cost of ERP software historically represents roughly 20% of the total project cost. Start the cost estimating process by getting two or three initial quotes for software packages. The packages should be within the software tier associated with your company size (see Chapter 5). Next, calculate the average package cost, and multiply this number by four or five (depending on how conservative you want to be) for the estimated total project cost.

The total estimated cost can be broken down into the major implementation cost categories by using some commonly observed percentages. These include:

- Consulting Services (PM, application, technical) = 50%
- ERP Software = 20%
- Application Development Services (programming) = 12%
- Hardware & System Software = 10%
- Education and Training = 8%

Project Funding Estimates

Undoubtedly, some type of project funding request is necessary before pur-chasing software and hiring consultants. Those who must approve the fund-ing will want to know the project schedule and cost. This will require more research and getting further into the specifics of the project.

In fact, one should also review the next section of this chapter (developing the baseline estimates) prior to completing the funding document, particularly the list of potential budget items.

When managed correctly, the bidding process for selecting ERP software and consultants can yield a much better understanding of the project. There-fore, gather as much input as possible from all ERP software vendors and consulting firms quoting the project, not just from those selected.

As outlined in Chapter 6, first inform all vendors that planning assumptions should accompany all quotes and that low-balling schedule or consulting cost will not necessarily result in winning the bid. This requires that each vendor get more into the project specifics. Go as far as to ask each vendor for a detailed project plan, and then take what you can get.

In addition, all vendors want the project to be funded. They will usually go out of their way to provide additional information beyond what they are selling. Of course, during the bidding process all of the above is free. Take advantage of this while it lasts.

Second, at this point the consulting cost estimates are mainly a by-product of the project timeline. Have each firm quote consulting hours by project phase in weekly buckets for each consultant over the course of the project. Beyond scope and other assumptions, this requires them to think through the activi-ties within each phase, the level of effort involved, and their responsibilities versus those of the client.

It is also important to work with all consulting firms to reconcile differ-ences in their quotes. If all firms are really quoting the same project, timeline and consulting costs should not vary significantly. This can (and should) be accomplished without sharing quote information between vendors. Inform the firms suspected of low-balling that they are underestimating the project. Next, let them make the adjustment they see fit, not simply by matching another firm's bid.

Also, remember that you (not the vendors) control what is submitted to management in the funding request. Even when insisting on valid quotes for consulting services, quotes can be so far off that adjustments upward of 50% are not unusual.

A good technique to estimate the overall timeline and budget is to start with the longest quoted schedule and cost (regardless of the firm selected) and compare against the guidelines previously mentioned for making early estimates. You might find it necessary to adjust upward.

Since at this point the ERP package and associated technology options are known, the IT group should be able to obtain quotes from third party vendors based on the technologies they will likely select. Early quotes for hardware and system software are usually in the ballpark (if there is a good estimate of the number of servers required).

For the project-funding request, always be conservative with the return on investment (ROI) since there is a human tendency to overstate benefits and underestimate costs. In the end, if the project cost is significantly underestimated, the project manager will get the blame.

Remember, the approved funding sets the *limit* on the total project budget, not individual line items; so provide some buffer. On the other side of the equation, it is my experience that senior management is usually more forgiving of projects that take longer than stated in the funding document. Therefore, the focus should be on the total cost.

Finally, if senior management has concerns about the project cost, by all means, revisit the numbers, but always change the project assumptions to reflect how the lower cost was justified. This might go as far as changing the project objectives and scope. If you cave now to unreasonable demands and reduce the costs without changing the assumptions, you will regret it later.

Baseline Estimates

Early estimates are necessary, but the project manager needs a handle to manage activities and expenditures on a daily basis. Unlike other plans or estimates prepared previously, the *baseline* schedule and budget represent the detail required to manage deliverables.

The baseline is also a tool to measure progress and actual costs for specific line items. This is why it is called the baseline since it is the yardstick by which all actuals or changes are measured.

The baseline schedule should be aggressive, yet achievable. It is prepared during the latter stages of the planning phase. At this point, we know much more about the project. The objectives, scope, and software rollout strategy are known. The implementation teams are now in place, and the software consultants are engaged to perform preliminary analysis and to assist with planning. In addition, senior management and project team education and software training and the *As-Is* analysis should be complete or well underway. In other words, for the first time, we have more information and all the project resources are in place to develop a real schedule and budget.

The major phases and deliverables outlined in Chapter 7 are a starting point for developing the schedule, plus sample or template ERP project plans are easy to obtain.

When using a consulting firm, project scheduling is an area in which their project management expertise should be leveraged to the hilt. One of these areas is how to use project scheduling software, when this knowledge does not exist internally. Finally, get educated on project scheduling concepts, principles, and tools when necessary.

The Detail Schedule

The right scheduling is a "top-down, bottom-up" approach, with several iterations. It is top-down in that it begins by *decomposing* (breaking down) the major phases and deliverables into specific tasks and task *dependencies* (i.e., Task A must complete before Task B can begin).

When breaking down the project work, at some point activities should be organized around each software module and business process. The reason is that current process analysis, the *To-Be* processes, configuration set up, testing, etc., are associated with modules and eventually business processes. It is best to plan the project the same way.

The scheduling process is bottom-up in the sense that the dates to start and complete major phases and deliverables are derived by the associated lower level tasks dependencies and *durations* (elapse time required to complete a task). Therefore, the project master schedule is not pulled out of thin air. It should be based on what must occur to complete each deliverable and thus the entire project.

Once the project management team constructs a first-cut schedule, they validate it for structural integrity. This ensures all tasks are included, dependencies are properly linked, resources are assigned, and task durations pass the test of reasonableness.

During validation of the schedule, pay special attention to the tasks on the *critical path*. This set of related tasks, when added together, determines the total project duration. For items on the critical path, it is prudent to revalidate tasks durations and dependences. Most project scheduling software will identify items on the critical path.

Next, determine the workloads placed on each team member by time-period and compare these to their committed project hours for the period. Adjust the schedule or task assignment accordingly. As imperfect as it is, most scheduling software contains some type of rough-cut capacity planning capabilities.

After working the schedule within the project management team, review the draft with the entire project team. This is an interim sanity check to gather more input prior to finalizing. It is an important step because those implementing the plan must believe and support it. It is best to get team "buy-in" early, rather than first presenting the schedule to the team after it is final. In the latter case, the project manager could become the only one committed to making it happen.

The Critical Path

Part of finalizing the schedule is to work the critical path to complete the project sooner. One way to compress (crash) the critical path is to shorten the task durations by adding more resources. Nevertheless, continuing to add more resources does not necessarily continue to reduce the time to complete a task proportionally. Eventually, the law of diminishing returns applies.

Another technique is to look for the *earliest possible start date* for items on the critical path. It may be possible to start some tasks sooner than the "hard" serial task dependencies suggested in the existing plan. Usually, there are critical activities that can run somewhat concurrently with others on the critical path, thus reducing the overall timeline. The best way to do this is to break up a single task into multiple tasks and then start the first one earlier.

For example, let us assume Task 2 can only start after Task 1 is complete (according to the current plan). Upon further analysis, Task 2 may really consist of Task 2A and 2B. It may be possible to run Task 2A concurrently with Task 1. Task 2B is now dependent on the completion of Task 1 and 2A. We have now taken the time for task 2A off the overall project timeline.

Finally, another technique is to decouple task dependencies for a few major items on the critical path and simply *force* them to start earlier. For example, software development is normally on the critical path (this is one reason to minimize software modifications). Nevertheless, it might be possible to get a jump on the design of a few high priority or difficult data conversions, interfaces, or software mods prior to the "official" design phase (of course, if the steering team approves the mod).

If you take this approach, treat each piece of major development work as a separate project, but linked with the overall project. These "mini-projects within the project" might have dedicated resources.

Of course, there are risks in starting something too early and encountering more rework than it was worth. The key is to find the "sweet spot"—when these types of tasks can start early without causing excessive rework.

Use of Slack Time

Once the critical path is set, another step in finalizing the plan involves using *slack time* effectively. Tasks not on the critical path have "slack." To a certain extent, it does not matter when these start or finish as long as completed before they become critical.

All ERP projects have slack activities, so use this time to your advantage by scheduling non-critical items when most convenient to the team, or to smooth the workload when resources appear overloaded.

Protect the Schedule

Murphy's Law says what can go wrong, will go wrong. Nevertheless, there is such a thing as being proactive to increase the likelihood that important events materialize as planned or to mitigate known project risk. The final step in building the schedule is to "protect" it by adding additional steps to help ensure that activities occur without major problems.

Often times, these additional steps are easy to identify and perform. For example, prior to an important software demonstration with stakeholders, do a quick "dry run" with only the team present to make sure the system and the team is prepared for the demo. Again, it is easy to identify proactive tasks to add to the plan by asking: "What can go wrong when performing this important task?"

Once the project management team is satisfied with the baseline schedule, a second review should occur with the rest of the project team to make any final adjustments prior to the project kick-off meeting.

Scheduling Mistakes

Building a schedule is always somewhat subjective, but the goal is to make it a good predictor of reality. Without attention to the items below, a schedule can be grossly understated.

1. **Inadequate Task Definition**

   When developing the plan, breaking project tasks down to extreme levels adds little value in the end. However, thinking through what must occur within each activity helps isolate certain steps that should be treated as separate tasks.

   A classic example of this is software development. Creating a single task for a software development project called "XYZ Programming" does not reveal the true scope of work. Developing software requires more than physically writing programs. It also involves analysis, specifications, and unit testing. In addition, these activities involve more than just the software developer. Most software development also requires the participation of the application consultant and functional analyst.

   Another example of poor task definition is failure to include tasks that are necessary to *transition* between projects phases (i.e., wrap up the work on one phase and prepare for the next phase). As an example, it is not realistic to assume the team can complete the first round of conference room pilot testing on Friday and immediately jump into round two testing on Monday.

   The final major pitfall in this area is not recognizing *planned rework*. Planned rework is different from unnecessary rework, since planned rework should be anticipated and can be of benefit to the project. However, unplanned rework is a project manger's worst enemy.

For example, the purpose of a design review is to gather feedback and input from stakeholders outside the team. Therefore, an expected outcome should be improvements in the system design considering the feedback. The process of updating the system design or set up to reflect the changes is rework, but it is "good" rework. Make sure you account for this type of rework in the schedule.

2. **Tasks Dependencies Are Not Well Defined**

   Almost every task should have at least one *predecessor* task (something that must complete first). Most project scheduling software identifies tasks with no predecessors.

   When it is difficult to identify a predecessor task for an item, the activity should be broken down further in order to build a definitive dependency relationship with other tasks in the schedule.

   Second, a task can have more than one predecessor. Since the accumulation of related tasks comprises the critical path, a single task not properly linked as a predecessor can result in an incorrect critical path. You do not want the critical path determined by simple mistakes when setting up task dependencies.

3. **Task Durations Assume Execution Is Flawless**

   The duration of each task is based on what needs to be accomplished, the estimated level of effort, and applied resources. Certainly, avoid building unnecessary fluff in the schedule. However, when implementing ERP, there are learning curves, unknowns, and meeting cycles, and as a result, not all decisions are made quickly. Account for these realities in task durations.

4. **Theoretical Versus Actual Hours**

   Even when the implementation team plans to allocate the agreed upon hours per week to the project, usually not all of this time is actually applied or 100% productive.

   Of course, always account for vacations and holidays when determining the amount of resources used for scheduling purposes. But it is not practical to include every conceivable item in the schedule or outside the project that consumes project resources. For example, there are project status meetings, coffee breaks, and sick days. In addition, team members at times will be pulled away from the project to address a crisis within the business.

As a result, for each resource, plan for less than the average number of hours per week theoretically committed to the project. This may seem trivial, but if resource availability is overstated by only 3-4 hours per week for each resource, this number can really add up over the course of a project. For this reason alone, the actual timeline could exceed the schedule by months.

The Master Schedule

The master schedule (i.e., schedule of deliverables) specifies the start and completion dates for the project, each phase, each deliverable, and who is responsible for the deliverable. This schedule is a summarized version of the detail schedule in that it only depicts the highest level of the plan.

Again, no detail schedule will unfold exactly as planned, but when done with the due diligence; minor omissions, errors, or inaccuracies at lower levels tend to cancel each other out at the highest level. In this case, the master schedule is still an accurate picture. This is what is important since the ultimate goal of any project manager is to hit the phase and deliverable start and completion dates, not necessarily each individual task in the detail plan.

This does not mean the detail plan becomes irrelevant once the master schedule is published. The detail is important later in planning weekly activities, assigning action items, gauging progress, making adjustments, and simulating the effect of proposed scope or resource changes.

The last step in preparing the plan is presenting it to the steering team for approval. The master schedule is typically the only schedule in which the steering team is concerned (although most executives want to know if there is detail to support it). In fact, having the detail as back up during sr. management's review of the master schedule reduces the chance that they will seriously challenge it.

The Final Budget

If the cost estimates in the project-funding document provided some room for the unknown, the final budget will probably not exceed those estimates. Thus, budgeting at this point should be a process of fine-tuning the numbers as more decisions are made during the planning phase.

When finalizing the budget, the biggest advantage is we now have a real project schedule with assigned resources. Better yet, the steering team and other major stakeholders have reviewed the schedule and they support it. This allows for revisiting the estimated number of hours per week for each consultant, and lock down the final consulting budget.

One way to fine-tune the consulting hours is to "ramp up and ramp down" the hours to better reflect how you plan to use each consultant at different stages of the project. For example, during the planning phase, project management consulting hours per week should be at their highest. Once the project is launched, the hours should ramp down considerably and stay relatively flat until the cutover phase begins.

How low one can go with project management consulting hours depends on what the project manager (assigned from within the company) is capable of doing, and the level of support required. Once the planning phase is complete, project management consulting could be perhaps one day per month or one day a week. If a PM consultant is required more than two days per week, one may question whether the client project manager is capable of performing the job.

In terms of application consulting hours, fewer hours should be planned for completing the preliminary analysis, the *As-Is* analysis, and when the project team is attending software training.

Preliminary analysis runs concurrently with the planning phase and this time is necessary for consultants to become oriented to your business and the project. However, it is not an efficient period during most projects since the project is still somewhat unstructured. So be careful not to overkill consultant hours during this time. A couple days per week during the preliminary analysis phase should be enough.

As discussed further in Chapter 17, most organizations can conduct the current process analysis on their own. An application consultant should be able to review the completed process maps, ask questions, and provide feedback without attending these meetings. This should take a consultant only a few hours versus setting through many meetings.

If the consultants are to conduct the team software training, this cost is part of the education and training budget. If the team attends training not provided by the consultants, during this time the consultants may not have anyone to work with or anything productive to do. If so, do not schedule them during this time.

During the prototype, design, and construction phases, application consulting hours progressively ramp up since more support is typically required during this time. How much it ramps up depends on how quickly the project team can learn the software. Once formal testing begins, consulting hours should again ramp down since testing should always be the primary responsibility of the organization.

Hours should increase for all types of consultants during the cutover phase so they can assist with final preparations. If the system is truly ready for go-live, consulting support after cutover should be heavy for only two weeks or so. Hours should then ramp down, and end completely within about four to five weeks.

In establishing a weekly budget for application consultants, as a rule 40 hours a week on a routine basis should be avoided since this means consultants are camping out on the project. On the other end, eight hours per week is not an efficient use of consulting time either. Time is wasted, as the consultant gets reoriented to the project every week. The hours scheduled for a consultant for most weeks should be zero or 16-32 hours.

Finally, the consulting budget is just a budget. The project manager will determine the *actual* schedule to be released to each consultant as the project progresses. This is important because if you give consultants all the budgeted hours, they will burn every minute of it whether it is necessary or not.

List of Budget Items

Figure 9 below lists budget items to consider at any stage of estimating:

Figure 9 – Potential Budget Items

EDUCATION AND TRAINING	SYSTEMS HARDWARE	CONSULTING
Sr. Management Education	ERP Servers	ERP Readiness Assessments
Industry Practices Education	3rd Party Bolt-on Application Servers	ERP Software Selection
ERP Software Project Team Training	Internal Network Upgrades	Industry Practices Consulting
IT Systems Administration Training	Telecom Service Upgrades	ERP Project Management
IT Application Development Training	PC's	ERP Application Consulting
Implementation Tools Training	Printers	ERP Software Technical Consulting
Books / Societies / Seminars	Data Collection Devices	3rd Party Systems Technical
APPLICATION SOFTWARE	Mobile Devices	3rd Party Bolt-on Applications
ERP Software Package	Storage Access Network (SAN)	Go-Live Readiness Assessment
3rd Party Bolt-on Applications		ERP Change Management

IMPLEMENTATION TOOLS	SYSTEMS SOFTWARE	CONTRACT PROGRAMMING
ERP Application Development Tools	ERP Package Proprietary Middleware	Data Conversions
ERP Application Set Up Tools	Operating Systems	Reports
3rd Party App Development Tools	Data Base Software	Interfaces
3rd Party Report Writers/Data Query	Web Services Software	Software Modifications
Project Scheduling Software	Client / PC Software	
Process Mapping Software	3rd Party Middleware / Integration	
Other Implementation Tools	EDI Software	
	Forms / PDF Management	**ANNUAL SUPPORT FEES**
	Job Scheduler	ERP Software Maintenance
	Data Back-Up / Recovery	All Systems Software Maintenance
		Hardware Maintenance
ADDITIONAL STAFFING	**OTHER**	Outsourced IT System Infrastructure
Project Team	Facilities and Equipment Upgrades	Outsourced ERP Application Support
IT Department	Travel and Living Expenses	ERP Software-as-a-Service
Temporary Staffing to Backfill	Hardware Shipping Cost	Implementation Tools
Temporary Staffing to Load Data	Project Team Incentive Bonus	
	Sales Tax	
	Contingency Factor	

Note: Some of the software and tools listed above may be included with the ERP software package.

CHAPTER 14

THE RIGHT TECHNOLOGY QUESTIONS TO ASK

The topic of this book is not technology, nor does an ERP project manager need to be a technical person. Technology alone is usually not the reason for an outright failure—unless one makes terrible technology choices.

The new system includes not only the ERP software, but also infrastructure products such as databases, operating systems, middleware, and hardware. These components can be from different vendors. Therefore, they must be carefully selected, installed properly, work well together, and be supported.

In addition, there are system administration functions within the ERP package that must be managed. Furthermore, at least some custom software development is usually necessary, and this falls under the IT umbrella. This necessitates an understanding of the software development tools that come with the package.

Keep Technology off the Critical Path

ERP presents a big enough challenge on the business side of the project, but allowing the technology implementation to become a bottleneck or constant headache only adds to these challenges. The project manager must at least understand the right technology questions to ask and have the technical leadership and resources in place to keep technical issues off the critical path.

No matter how great the ERP software functionality, an unreliable system riddled with bugs or inadequate support can significantly delay a project or cause excessive system downtime and slow transaction processing.

These potential issues are not just isolated to internally hosted systems. An outsourced system can bring a business to its knees just as quickly as any internal system. When it comes to technology and support, the out-of-sight out-of-mind mentality can get you in trouble. Therefore, most topics in this chapter are also valid for SaaS, Cloud Computing or any outsourced system.

The "Bleeding Edge" Is a Choice

In Chapter 5, we addressed the need to stay away from the bleeding edge of technology. One way to avoid it is to choose a package that can utilize mainstream technology products. This philosophy is best even when it is necessary to move away from technologies familiar to the IT group or deny you the "latest and greatest" stuff.

When higher risk choices are necessary for good business reasons, recognize what you are getting into and bake in plenty of extra schedule and budget due to potentially more technical consulting, testing, and system issues.

IT Department Linkage

For many different reasons, the IT department may not be properly aligned to support the project. One thing is for sure: If there are any longstanding political battles between your IT group and other areas of the organization, now is the time to address this issue.

IT linkage into the project is important since technical input is necessary and someone must manage the hardware and technical software installations. In addition, the project team requires technical support during the project and so do the end-users once the system is in production.

Worth repeating here: The leader of the IT department (CIO or IT Director) should be on the executive steering team. An IT manager should be assigned to the project management team and report to the project manager. A technical support team should be part of the project organization and a software developer should be a member of one or more application teams. In addition, most projects will need technical expertise from the ERP software vendor, consultants, and third-party providers.

The Right IT Skills

Beyond IT leadership, the other skills to assess are those of the system administrators, application developers, and technicians that must keep the new systems running everyday. IT training and consulting support are always necessary in these areas, but do not assume that every person in the IT group can successfully make the transition to new tools and technologies.

Some IT department employees may not have the aptitude to learn the new stuff and others may not have the desire. When IT staffing changes are necessary, it is best to anticipate them and to make the moves as early as possible.

Finally, when replacing older applications that can only be changed by modifying the software, a new skill in IT is knowledge of how to configure the new system. Often those in the project roles of in-house application consultant or functional analyst transition into permanent positions in the IT department to fulfill this need once the project is complete.

The IT Bandwidth

For each IT role, determine if the number of people currently available is enough. When short on resources and skills, do not hesitate to augment the IT staff with outside support.

Keep in mind, technical consulting (which is different from software development or application consulting) is usually the smallest expenditure in the consulting budget. Any additional costs to get the system infrastructure set up correctly are unlikely to bust the budget.

In addition, when mainstream or non-proprietary technical products are selected this is an area where finding qualified outside resources for a competitive price is not difficult.

The application development skill is usually specific to the software package. Typically, the primary consulting firm engaged with the project also provides software development support. But this is another area where it is possible to find good and less expensive resources from a third-party vendor (particularly when using a Tier 1 or 2 package).

Estimate the total number of application developers for budgeting purposes. In doing so, the spike in the application development workload tends to come at the end of the design phase and runs through conference room pilot testing. Just prior to this time, begin to line up the development resources. If there are not enough resources to cover this peak in development work then the project can fall behind schedule very quickly. As always, do not bring in outside developers until there is a clear and agreed upon definition of the programs to be written.

Finally, legacy system support will continue during the project, and keeping current systems running is always the number one priority of any IT department. This will affect IT resource availability for the project. The issue of legacy system support can be partially offset by reassigning support responsibilities and placing a moratorium on current system changes (i.e., no new IT projects on the old system unless business-critical and approved by senior management).

IT Knowledge Transfer

In order to maintain a reliable, available, and high performance system, the IT group requires knowledge. Training is usually necessary in areas such as the application development tools, system administration, security, and third party software such as EDI, job schedulers, databases, and operating systems.

Beyond formal training, like anyone else, IT employees learn by doing. When technical consultants are engaged to perform software installs or system administration activities, the IT group should work closely with them and perform most of the hands-on tasks under the direction of the consultant.

Immediately after formal training, programmers can cut their teeth on reports and simple screen changes, gradually moving into more advanced development. In addition, writing some useful applications for the project team with the new programming tools is a great way to expedite the learning curves. For example, the team will need a database to track project issues. Instead of using spreadsheets or buying software for this purpose, develop it in the new software.

Software Instances

Most projects will require more than one instance of the software (system environment) such as one each for application development, testing, and the live production system. Also, there could be a separate instance for training and one for each phase of testing.

When multiple business units will use the software, most IT people prefer not to run a separate occurrence of the live system for each business (unless absolutely necessary). When faced with this decision, think it through carefully since a wrong choice here is difficult to reverse once the system is in production.

Many ERP packages have the flexibility and functionality to support multi-businesses within a single database. Having all businesses operate within the same environment is always best from an IT cost standpoint since it usually means less systems infrastructure to purchase and support.

Of course, one of the advantages of ERP is the enterprise integration provided within a single database, and this is lost with multiple instances. If separate environments for business units are necessary, and these businesses must integrate at some point, then interfaces must be developed.

There are a few examples when a separate software instance for each business is the best approach. Some companies have a philosophy of "business unit separation" for strategic or other reasons. There is a belief that each business unit must stand on its own and each should have the autonomy to be successful. In this case, the tangible benefits of separation from a business flexibility standpoint can outweigh the additional IT costs associated with implementation and support of multiple systems.

Another scenario when separate environments can be the best choice is when each business unit will use the same software modules, but each has very different requirements (to the extent they represent entirely different types of businesses). Moreover, there is little or no integration required between these businesses.

In this example, it may be difficult to squeeze more than one business into the same software environment due to limited fields, multiple field usages, and security limitations or complexities. Finally, many multi-business ERP systems have some software modules or functionality that is not easily separated by business. In certain areas of the software, the business units might step all over each other when attempting to work within the same system.

Technology Rollout

Information technology planning and the evaluation of technology options should begin before an ERP package is selected. Once purchased, final technology choices, IT training, and the installation of the infrastructure should start immediately.

The technology plan to support the project includes major tasks, schedules, and responsibilities for installation of all hardware and software to run the entire ERP infrastructure. This infrastructure must be available, supported, and working smoothly early in the project so not to delay the project team.

The Package Install

In addition to third-party products, any ERP package has its own set of system components. Physically installing the ERP software on the servers, creating environments, and other IT setup activities are critical. When not done correctly, problems are usually immediate and systemic in nature. In some cases, major issues are discovered later, necessitating a complete reinstall and this can really slowdown a project.

I always recommend utilizing the software vendor for the physical install of the ERP package on the servers. This is the case even when the software vendor is not the primary consulting firm hired for the project.

First, the software vendor wrote the package and usually has more technical resources and expertise to address any issues during the install. When using your consulting firm for the install, if there are major problems encountered, there could be delays in gaining access to the right technical people within the software vendor organization. Also, there is a greater chance of vendors getting into a blame game in this situation.

Before the technical installation, ask the consultant (who will do the work) for a specific list of prerequisite tasks the IT department should complete before the engagement. This helps avoid delays and unnecessary consulting costs.

After a technical consultant from the software vendor installs the package, on-going support can be turned over to the primary consulting firm or the IT group (with the help of the primary consulting firm) for the duration of the project.

System Performance

If there was diligence in selecting the package and technology products, the performance of the system should be a non-issue. However, the project manager should be proactive to ensure that the systems supporting the ERP package are properly sized and tuned to avoid performance headaches later.

When purchasing servers, remember hardware is relatively inexpensive, so provide plenty of headroom to cover estimating errors and some future growth. Also, size and tune the operating systems and database to maximize performance. Finding a qualified vendor for this tuning is not difficult, because many organizations use these same products.

As mentioned, the ERP systems infrastructure consists of products from many vendors that must play well together. From this point forward, it is about fixing technical bugs and addressing any system bottlenecks that arise.

Major system performance issues typically relate to how the ERP package interacts with the database product or due to network or the telecommunications bandwidth.

System performance should be monitored closely during the conference room pilot and training phases to spot potential bottlenecks. These are the times prior to go-live when there will be the greatest load placed on the system. Furthermore, I always recommend some amount of volume testing prior to go-live.

IT Policies and Procedures

The IT group must have policies and procedures in place to maintain a stable system, to support the efforts of the project team, and support the business after cutover.

This includes policies and procedures to manage security, computer operations, software change control, and for moving programs, files, and data between system environments.

It is also important to understand how the project team will effectively communicate with the software developers. If the goal is to minimize rework, realize that custom programming can create major rework when there is no standard for documenting system requirements. This occurs when everyone on the team has a different method to communicate programming needs to software developers, resulting in incomplete or misinterpreted information.

When interacting with software developers, there should be a standard *programming specification* format for the different types of development (mods, data conversions, interfaces, and reports). Also, once custom development is approved, we must ensure that stakeholders are in consensus on the system requirements before programming begins.

Normally, the software vendor and/or consultants can provide examples of design and program specification formats. Take all available examples and simplify the forms when necessary. As always, the important point is that the documentation communicates, not to create mounds of paperwork or unnecessary barriers to accomplishing the work.

CHAPTER 15
PUSHING AND PULLING FOR CHANGE

Do You Need an Expensive "Program"?

Many software consulting firms love to sell their clients big expensive change management programs during an ERP implementation. Probably next to readiness assessments, change management is the motherload of all ERP add-on work for the software consulting industry.

Actually, readiness assessments are used as tools to convince their clients that they need a change management program. Therefore, consulting firms get a double add-on beyond the normal tasks of helping to implement the software.

Some consulting firms focus entirely on change management and get their foot in the door at the very beginning of the project. While it is possible some organizations need help in this endeavor, personally, I am a little baffled for three reasons:

1. ***What secrets are consultants telling their clients about change management that the client should not already know?***
   Managing change is about understanding human behavior. We should all know something about it because we are all *human*. It is not as mysterious as many consultants want you to believe.

2. ***What are you paying your existing consultants to do?***
   Change management activities should be just another part of any ERP implementation process. It is what the organization and existing consultants on the project should already be doing. If your software consultants do not add value in this area, why not? In this case, you might have the wrong consultants.

   Similar to managing project issues and technology, change management is a project management *thread* that runs through the project cycle. It is not something you just go out and do one day. Many project activities and responsibilities to implement the software have major change management implications—if the value is recognized and when done correctly.

3. ***Why so many change management consultants?***
A consultant should coach, but the organization should be responsible for most tasks associated with managing change. In fact, many consultants running around managing change may ignite more resistance to change!

The Grapevine

Whenever a major change is brewing, it is human nature to assume the worst until you hear something different. Think about it, if some big nebulous change were heading your way at work, what are the important questions you might have?

If you cannot get clear answers from those initiating the change, you naturally turn to another source where information flows freely. This pipeline of information is the informal "grapevine" and every organization has one.

Of course, the grapevine is full of half-truths, naysayers, and even lies, but things really heat up when a big change like ERP is underway. Never underestimate the power of the negativity coming from the grapevine. The groundswell of resistance can get so intense it can literally bring down a project.

Even if the grapevine does not derail the project, it wreaks havoc in terms of additional time and costs, possibly affecting the quality of the implementation. With many employees fighting the project and others running away from it, there are not many left that are eager to help. Worse yet, over time, even the project team can turn sour. When this occurs, the project manager is hung out to dry—on the grapevine.

Managing Communication

If the ERP project manager fails to control communication to the masses, the grapevine will. We must stay one-step ahead of the naysayers to defuse any negativity or misinformation.

This requires a communication plan with responsibilities, dates, and target audiences. We need to get out the project facts (early and often) to dispel the fiction by answering the basic questions all affected employees will have:

1. What specifically is the change?
2. Why is the change important to the organization?
3. When will the change occur?
4. Who is working on the change?
5. Will I have input or be involved with the change?
6. When will I be trained on the system?
7. How will the change affect my department?
8. How will the change affect the way I accomplish my job?

In fact, failure to answer these basic questions is a disservice to employees. Most want to understand new expectations because they want to do a good job.

The first two communication items above are the responsibility of the executive sponsor. The project manager is responsible for questions three through six. All of this should be communicated across the organizations during and immediately after the project kick-off. Each application team addresses the last items usually during the design phase and end-user training.

All too often communication and change management activities start with a bang at the beginning of the project, but the effort fizzles out shortly thereafter. In other cases, there is very little change management and the project team mistakenly expects the end-users to embrace the system during training. The key is to start communication and change management early, and maintain it throughout the project or potentially face heavy resistance to the new system at some point.

Pushing for the Change

Managing change is much more than timely communication. There are three additional key elements.

If senior management does not push the change, it may not happen. Early on, some employees will not want to easily give up what is familiar (not matter how bad it is) to embrace something yet to be defined. In addition, some managers will fight the change because they feel a loss of control or the project threatens their kingdoms. Others simply resist any type of change.

In order to push the change, the executive sponsor must get involved. As said before, if the project is not perceived as important to senior management, it will not be important to anyone else.

The biggest job of the steering team is to build a sense of accountability for project success within the middle management ranks. Success is not just the responsibility of the project team.

As discussed before, new performance measurements for process owners and functional managers go a long way in pushing the change and sustaining it. Measurements also serve as a focal point for continuous improvement.

By implementing new expectations (and measuring them), the message is clear: Senior management is serious. Financial incentives for achieving performance goals should be considered.

Pulling for the Change

We must get the majority of employees *pulling* for the change because they see the benefit in performing their daily jobs. This is more than attempting to sell software as the solution. It is hard to sell a bald man a comb, but it does not hurt to try!

Most employees will pull for the change when solutions are actually better than what they do today. The solution works from them and for the company. What many fail to understand is most employees want to be more effective and contribute to company success.

In order to develop truly better solutions we need a project team with business analysis skills and who are change agents and knowledge workers. Equally important, the project team must educate key stakeholders on the software and involve them with decisions that directly affect them.

This is the only way any solution will fly in the real world since it creates more user ownership in the solution. In fact, process owners and key users are the only ones that can make the change stick after the project team disbands.

In addition, when there is ownership these same employees will probably sell the change to other employees who might otherwise resist it. What a major twist: The grapevine is now filled with good news!

The Naysayers

The final component of change management is very simple: Deal with the perpetual naysayers, one way or another.

It is very important to recognize the difference between a naysayer and someone with legitimate business concerns. As a project manager, legitimate issues should be welcomed because the project cannot be successful by simply ignoring them. Furthermore, some people need time to absorb the implications of the change before accepting it.

It is worth noting, when other components of change management are implemented correctly, the great majority of employees will support the project, have legitimate concerns, or become somewhat indifferent. None of these hurt a project, and there will be fewer naysayers.

On the other hand, the last thing we want are employees openly or subtly attempting to sabotage the project, erecting unnecessary roadblocks, or spooking other employees. This is potentially the problem with some in management positions and employee attitudes toward change can reflect those of their boss.

If the naysayers are not addressed early, they can quickly turn other employees into naysayers. In fact, removing a few people barriers will silence other naysayers and make those on the fence think twice about becoming one.

CHAPTER 16
KNOWLEDGE TRANSFER: RAISING THE BAR

The Promise of Knowledge Transfer

Most ERP projects are filled with the promise that consultants will transfer software knowledge to their client. Yet in many cases, once a project is over, the project team is clueless with how to make software configuration changes and may even struggle with performing basic transactions in the system. So, what gives?

First, there never was a real strategy to make it more than just a promise. Second, when push comes to shove, this once important concept of "learning" suddenly becomes something we worry about later. Later, of course, never happens.

This is similar to consultants building a spaceship to take you to Mars with the understanding that we will not plan the return trip until after you get there. There are real consequences for assuming that software knowledge will automatically cross-pollinate. These include consulting cost overruns, sub-optimization, and an inability to leverage the software investment once the consultants leave. More specifically, some of the consequences are:

- **Not Invented Here Syndrome.** When there is a lack of understanding of industry best practices and the software, there is a tendency to reinvent the "solutions" wheel.

- **Paying expensive consultants to perform tasks that the client could (and should) be doing.** Forget about owning your project. Without a certain level of software knowledge, it will be difficult for the organization to fulfill even its most basic project responsibilities.

- **Fostering resistance to change.** Many times, we are our own worst enemy. It is not difficult to imagine why employees refuse to buy into something they do not understand, no matter how great it sounds. When consultants are running the show (because the project team has not learned a thing), understandably, many employees will view the project as a disaster waiting to happen.

- **Poor software quality (not just a perception).** Inadequate software knowledge creates a barrier to involvement when designing the software and new business processes. Unless the consultants know everything (which is never the case), something important will fall between the cracks. Alternatively, an organization that understands the software capabilities and configuration settings can at least ask the right questions, spot things that are wrong, and perhaps even take the lead in developing solutions.

- **Paying consultants to camp out for years after the initial go-live.** Someone must hold the hands of untrained users and make simple software changes required by the business. In addition, eventually there will be requests to utilize additional features and modules within the system that were not enabled within the original scope. Who will implement these projects? A similar issue exists with regard to the ability to implement new releases without an army of consultants.

- **When hiring consultants cost too much, the software remains static while the business needs change.** The reasons are:
  - No one internally is aware of what the software is capable of doing.
  - No one internally understands how to make the changes.
  - In the meantime, users develop procedures to work-around issues inherently addressed by the system, perhaps with simple software set up changes. We are now back to the sub-optimization that existed with the old system!

- **Once the system is up and running, software modifications are performed to address requirements already supported within the standard package.** Unbelievably, this happens all the time.

- **As employees change jobs or leave the company, or new ones are hired, consistency in procedures and knowledge of the system slowly erodes.** This is inevitable since no one can explain the big picture or knows the original intent of the software design. Additionally, the work procedures are outdated, and there are no consistent resources available to perform training. For lack of a better choice, users must fend for themselves.

- **After several years of struggling with the software, the company finally hires an employee from the outside with the expertise to provide needed support.** Unfortunately, had they focused on knowledge transfer from the start, a lot of grief could have been avoided—not to mention the time and money that could have been saved.

- **By now, some have recommended outsourcing support of the application with the hope that the problem goes away.** Don't kid yourself. If no one internally can make sense of user requests (from a software design standpoint) or, better yet, make the configuration changes, outsourcing may cost much more than originally envisioned (change orders).

Knowledge Transfer Disconnects

In order to understand why such an important aspect of an ERP implementation often goes astray, it is necessary to discuss a few disconnects with training and knowledge transfer in general.

- **The emphasis is on the budget.** For example, many consultants say expenditures for training the project team and users should be approximately 8% of the ERP budget. This sounds about right. However, knowledge transfer cannot be measured by how much we spend or how many people attend training. It is about planning and execution—using the budget wisely.

- **Training is perceived as "fluff".** Many managers view training as just a nice gesture to employees. They might ask, "What is the return on investment, or is it just fluff?" It depends on the definition of fluff, but lack of software knowledge within the company is a major contributor to the age-old problems: ERP takes too long, cost too much, and the benefits never materialize.

But for those searching for the hard numbers, a reduction of up to 70% of the application consulting budget is not unheard of (and this does not consider the consulting cost savings over the life of the software).

- **Cutting corners at the expense of learning.** The company is investing a significant amount of time and money in implementing a sophisticated tool like ERP. Are we now saying it does not matter if anyone understands how to use it?

- **Acquiring software knowledge is viewed as an event.** For example, send the project team off to software training, and they return, magically transformed into software experts! Formal up-front team training is a necessity since it lays a solid foundation, but it is just the beginning of the learning process.

- **Not understanding the difference between education and software training.** Most ERP software packages are designed around industry accepted business practices, operating philosophies, and techniques that have evolved over many years. Today the issue is not whether there is a body of knowledge with supporting software tools, the issue is whether organizations are educated enough to see the value or understand the proper use of the tools.

First, many have lost sight of the fact that there is a difference between education and software training. End-user software training is very important, but it is mainly about how to perform transactions in the system and how these transactions related to your business processes. This can be very different from understanding the original intention of the software design in order to apply the tools properly.

The issue of education vs. software training is analogous to training someone to fly a Dreamliner airplane but not explaining the concepts of jet propulsion or flight (trained but not educated). The opposite case is explaining how to use a chainsaw but not training on the best way to cut down the big tree.

Both of these are scary propositions, but in terms of ERP, it is about failure to achieve the benefits. The root of the problem is that managers and end-users never changed their behaviors to take advantage of the tools, and a big part of this is lack of education.

Software Knowledge As a Deliverable

The transfer of software knowledge is a process of discovery involving *cycles of learning* that run throughout the project. For example, the typical training plan developed by consultants only addresses initial project team training and user training, but nothing in-between. Interesting enough, the "in-between part" requires the most planning and provides the most opportunity for learning!

Similar to other project management threads discussed previously, knowledge transfer requires a strategy and a focus. The new paradigm is this: Knowledge transfer is a deliverable, not something we just hope materializes.

A deliverable is not a deliverable without expectations and responsibilities (for both the consultants and the project team). We must make it clear that transfer of knowledge is a priority and progress will be measured.

At the same time, learning is a two-way street. The consultants cannot spoon-feed the company forever. Ultimately, the organization is responsible for learning the software.

There should be activities within the implementation process to facilitate and reinforce the transfer of knowledge. The goal is to increase the cycles of learning to crest the learning curves so the knowledge can be applied to the project as soon as possible.

The Knowledge To Transfer

In addition to senior management education and IT training, project team knowledge is also critical. If the team does not understand how to use the software, do not expect anyone else in the organization to understand it either.

Once the team is more comfortable with the system, key end-users come into the fold—first, the power users, and then others. This occurs prior to training all users since these individuals play larger roles during the project.

As mentioned, the remaining end-users (the masses) are trained just prior to cutover to the new system.

The following are the important areas of knowledge that should be acquired and the primary audience for each:

- **Industry accepted business practices (as necessary).**
  - Project Manager, "In-house" Application Consultant, Functional Analyst, Team Leader, Steering Team, Key Stakeholders

- **The ERP package capabilities, features, and functions.**
  - Project Manager, "In-house" Application Consultant, Functional Analyst, Team Leader, IT Analyst, Power Users

- **How to configure the software functionality within each module.**
  - "In-house" Application Consultant, Functional Analyst, IT Analyst

- **How to use the software to perform specific job functions within the company.**
  - Team Leader, Functional Analyst, "In-house" Application Consultant, Power Users, remaining End Users

Knowledge Transfer Tips By Project Phase

The following is a list of actions and considerations to ensure knowledge transfer occurs during each project phase. Also, see Figure 10 at the end of this chapter.

Preparation Phase

Industry Practices: More Than Just a Seminar

As mentioned, there is a difference between education and training. ERP education is more than just a seminar of "best practices" put on by a software vendor (with a hidden agenda) in a room full of 300 other clients. It is about getting some real, independent education, not focused on any specific package.

The great Joseph Orlicky wrote a groundbreaking book titled *Material Requirements Planning* (a forerunner to ERP) back in 1975. It addressed how to plan and control materials in a manufacturing plant. The most interesting thing is that Mr. Orlicky (an employee of IBM) never glorified the role of software. He believed that if one did not understand the underlying concepts, principles, and practices of MRP, the computer was not much help anyway.

Orlickly and Oliver Wight went on to build and grow the "religion" called APICS (The American Production and Inventory Control Society), the best thing that ever happened to manufacturing practitioners. Every industry or discipline has a great religion, so become a scholar of yours.

There are practitioner-based societies dedicated to the advancement of knowledge in many areas and some offer certification programs. Team members should be encouraged or required to take these education and training courses before the project gets started. Better yet, they should become certified in an applicable discipline. These are the first steps in developing a few *knowledge workers*.

Consultants as Teachers

In Chapter 6, we discussed selecting the right application consultants. Beyond just a good understanding of the software and industry, here are some of the intangible skills that enhance their ability to transfer knowledge to their clients:

- **Consultants who can communicate.**

   A consultant with software knowledge is one thing, but if the consultant is a poor communicator, it undermines the transfer of knowledge. In addition, a consultant that does not say much may also not know much.

- **Consultants who can delegate.**

   With guidance from a knowledgeable person, most employees learn by doing, struggling a bit, and making a few mistakes. Moreover, seemingly insignificant consultant behaviors can create barriers to project team learning. These behaviors include:
   - *Consultants that always want to "drive" (not allowing their client to touch the keyboard).* Chinese Proverb: I hear, and I forget; I see, and I remember; I do, and I understand.
   - *Consultants that refuse to delegate.* The team should be required to work on its own at times with some meaningful outcome.

As they become more knowledgeable, the consultant should delegate some of his or her tasks to them, and then review their work.

Planning Phase

Strategies That Can Inhibit Knowledge Transfer

Rapid Deployment and fixed price projects may or may not result in a faster or less expensive implementation, but are definitely more consultant intense or date driven. This means the consultants may not have the time for the transfer of software knowledge.

The Desire and Ability to Learn

As mentioned before, staff the project team with employees that are ready, willing, and able to learn the software. Also, no consultant can educate a client that is not available. Management must allow the proper amount of time for the team to participate and there must be a low tolerance for meeting "no shows" (for this and for many other reasons).

The Knowledge Transfer Scorecard

As part of controlling the project, later we address the need for the project manager to schedule periodic meetings with each application team. One agenda item for these meetings is to review the progress of learning.

First, work with the consultant to define the software knowledge topics for each module and create a scorecard for the application team. Prior to meeting with the project manager, the team (with the help of their application consultant) performs a self-assessment of learning progress for each topic on the scorecard including knowledge gaps. During the meeting with the project manager, develop get-well plans to address the knowledge gaps.

Start these periodic meetings soon after initial project team training, because if the team falls way behind the learning curves, it will be very difficult to catch up later. Finally, like other critical success factors, learning progress should be on the agenda of executive steering team meetings.

Plan to Transition Consultants Off the Project

Many consultants say one of their goals is to work them themselves off the project by making their client self-sufficient. In spite of this good intention, the project manager must plan for this to occur if it is actually going to happen.

Based on project milestones, develop an aggressive, yet realistic schedule to transition outside consultants into purely support roles, particularly in the application area. As the team begins to get up to speed and confidence grows, there should be a pre-defined point when primary responsibility for the application shifts from the consultants to the team. Beyond this point, consultants are used less, playing coaching and supporting roles, and eventually are involved only when necessary.

This transition should be reflected in the responsibility for deliverables at each stage of the project and, to a certain degree, in the consulting budget. However, the consulting budget must be reasonable in case you need the hours, but you can always give the consultants fewer hours than budgeted.

Depending on the skills of the team, it is not unheard of to begin this transition after the design phase. At the very latest, the organization should take primary responsibility for the application once formal testing begins.

Plan for Just-in-Time End-User Training

It is not surprising that poorly executed end-user training results in a user community that does not know how to use the system. Starting this training much more than two weeks prior to cutover is generally a problem since most users will forget what was learned. On the other hand, the more one compresses the timeline for just-in-time training, the higher the demand for training resources over a short period. This must be considered during the planning phase of the project.

Discovery Phase

Do "Formal" Project Team Training

Project team training should conclude before the prototyping phase begins. When done correctly, traditional classroom style training is of considerable value since it lays a solid foundation for continued learning.

Over the last few years, many consultants advocate skipping formal classroom training for the team and instead doing it "on the fly" during the project. No doubt, informal knowledge transfer is part of the learning process, but relying on this alone has some major drawbacks.

The recent push for informal training as the only method began with the rise of the rapid deployment philosophy. But it is also an indication that many do not understand the purpose of project team training. This training should be focused entirely on learning the software navigation, capabilities, and basic transaction processing.

In addition, I believe the notion of forgoing formal training reflects past failures to plan and execute it properly. When not done correctly, formal training is a colossal waste of time and money. But it does not have to be this way.

Formal training is defined as instructor led, in a classroom setting, with a pre-defined lesson plan, a pre-configured training system, training scripts, supporting data, and other materials. The training documentation received should be for the current release of the software (and is of considerable value later).

Formal training is developed be people that not only understand the software, but also have expertise in teaching methods. In addition, it is delivered right out of the box. It is most effective when conducted by a professional trainer that has taught the same class many times before. Previous consulting experience in the application area is also important for any trainer in order to answer student questions that are "off script."

Formal training, when compared to informal methods, usually provides a broader perspective of the software capabilities, which is exactly what is needed at this stage of the project. In addition, these courses are time tested—not developed from scratch for a single training event and not highly customized. This usually results in a more complete coverage of the software capabilities, fewer disruptions (due to training system setup issues), and a more consistent and smoother delivery.

I prefer courses offered by the software vendor or a third-party training resource specializing in the package, since most can deliver all the above. Often the issue with consultant-developed training is that they do not have access to all the resources to conduct formal training as described.

Second, an application consultant may understand the software very well, but may not be the best at structuring and delivering the software knowledge in a classroom setting. One reason is most of their experience is informal training.

The next issue is the need to reinvent the training wheel since usually consultant delivered courses must be developed from scratch—including a lesson plan, materials, setting up the training system, and loading data to support the training. In the end, this may cost more and deliver less than pre-designed training options. Not to mention there is a greater risk of encountering software setup bugs during the training causing class disruptions.

Amplifying these issues, many software consultants are overly accommodating with their clients regarding how far they customize project team training. Many go well beyond what is reasonable or advisable.

For example, some consultants allow the client to make too many assumptions about what software functionality to exclude from this training. If something is obviously not applicable to your business or clearly not within the project scope, by all means, skip it. But remember, you may use capabilities that at first appeared not useful or relevant.

Finally, some consultants customize training to the extent that it is no longer training, but instead a premature design or testing session. Again, the goal of project team training is to learn the software capabilities and basic transaction processing—not to design or test the system. No one at this stage of the project has enough information to design or test anything.

The Training Courses to Get Started

In selecting courses, remember not everything can be learned in a classroom, no matter how many classes you take. Some modules may have multiple course offerings addressing progressively more advanced software functionality. However, the basic courses for each module are usually enough to get learning on the right path. Also, courses on how to configure the software should be taken immediately as well. This knowledge will be used early in the project.

After the team has worked with the system for a while (with the assistance of consultants), take the advanced courses, if necessary. With this approach, the student has the chance to absorb the information and is better prepared to take the advanced topics with a list of specific questions.

Take the Configuration Classes

Classes to learn how to configure the software are very important if the plan is to eventually wean the company from consultants. This knowledge can pay big dividends in the long run.

Many companies are beginning to realize the importance of these classes. Forty-two percent of respondents to a recent survey by Klee Associates (a SAP and JDE training and consulting company) cited configuration-level training as a major training need.

When available, these courses get into the parameters, switches, and settings that enable certain software capabilities to support the needs of the business. While most classes focus primarily on end-user transactions, the configuration classes address what is behind the curtain, though software programming experience is not a prerequisite for these courses.

The list of attendees for configuration classes should be limited to the application team. Though many others provide input into the setup of the software and must understand how the system works, they do not require knowledge of how to configure the software.

Right-Size Attendance for Project Team Training

Do not send a cast of thousands to project team training. A common mistake is to include many managers and end-users (not on the core team) that eventually must know something about the system.

Project team training is not about specifically how the software will be setup and applied within the business. This is often a misperception of users who want to attend. The intention of project team training is to understand the system capabilities, basic transaction processing, and configuration settings to prepare the team for the project.

Therefore, to get the biggest bang for the buck and to use employee time effectively, project team training should be limited to the project team (and perhaps a few exceptions discussed later).

In addition, each application team should only attend the training for the module in which they are directly responsible. Similar to training users, each team has the responsibility to share their knowledge with other teams at the right time. This also provides a greater incentive for them to get the most out of the training sessions.

This generally means project team training for each application team includes the team lead, functional analyst, in-house application consultant, and the IT support analyst. If the project manager wants to attend application training, no problem, but on a large project there might be a better use of time.

The last exception could be a few important process owners (key stakeholders) that are recognized in the project organization chart. If they want to attend project team training, do not stand in their way since you need them on your side. But explain to them this training is not about specifically how the software will be used.

If the formal training is to be conducted at your facility or for courses that are web-based, there is one approach to include some employees not on the team without busting the training budget. Negotiate with the vendor to allow a few additional attendees that are "observers" only and are free of charge. Observers do no hands-on training, but simply watch and listen from the back of the room. This may satisfy those that want to attend but have no immediate need to know.

The Right Training Delivery Method

Project team training can come in many forms such as on-line self-paced, on-line instructor-led, instructor-led at the vendor site, or at a company facility.

Though usually less expensive, I recommend on-line courses only for basic navigation or foundational/common topics. For the core or more advanced modules, face-to-face interaction with the instructor and others in the class is a considerable advantage.

The decision to train at a company location (not the vendor location) is usually driven by cost. But removing the team from the daily routines (and distractions) to focus on learning is a major benefit of training at the vendor facility.

Scheduling Project Team Training

Properly scheduling project team training is important. The goal is do it *just-in-time*. The time available for team training runs concurrently with planning and the *As-Is* process analysis, and is completed just before prototyping begins.

When scheduling training, work backwards from the start of prototyping considering the required sequence of course offerings (prerequisites). That is, delay the start until as late as possible. As stated previously, completing training too early is a mistake since once it is complete, the team must use the new knowledge or lose it.

However, scheduling training can be a balancing act especially when individuals must attend multiple courses. While we want to schedule courses as late as possible, scheduling courses immediately after the other is like trying to drink from a fire hose. For those attending, there is too much information coming too fast.

The best method is to stagger the courses within the training window mentioned above. This provides some time to absorb and reinforce what was learned through self-help and other training follow-up activities (discussed next).

This is not to say developing a good schedule for project team training is easy, particularly when working with pre-defined course schedules published by a vendor. At the same time, many vendors are flexible regarding how and when training is available to support customer needs, even for training at their location.

The "Playground" Environment

When the team returns from training, it is best to have a system environment immediately available for self-help and follow-up learning. This includes revisiting the training materials, exercising the training scripts again, and exploring the software further.

Taking advantage of this playground or "sandbox" should not be optional, but mandatory for those returning from training. Some consulting support may be necessary for further learning in the sandbox, but make it an extension of formal training, not the beginning of prototyping or system design.

If nothing else, the playground environment can be a stand-alone PC "demo" version of the ERP software (with demo data) that is used until the test environment is ready and prototyping begins.

Accelerate Knowledge Transfer

Like project team training, prototyping is part of discovery so do not cut it short. Beyond being a prerequisite to developing a solid system design, prototyping accelerates knowledge transfer early in the project. Equally important, the application team should perform the initial software configuration to support prototyping based on business needs, but with the knowledge and support of the consultant.

Design Phase

Enter the Power Users

Power users and key stakeholders are a select few from the user community designated to get more involved with the project at the right time. As they enter the picture, the team trains them on the software.

This training is hands-on, but less formal than project team training. It also focuses on how the software *could* be used in the business (not necessarily all capabilities). This training should begin during the design phase so power users and stakeholders are in a better position to provide input and validate design proposals.

Technically, this is the first group of end-users (outside the core team) to receive training on the new system. It is also the first time the project team are trainers, not just learners. In order to train, one must know something about the software. Hence, the team's ability to train the power users and stakeholders is an important knowledge checkpoint.

Get the White Papers

During the design phase, the team gets further into the details of the software setup to validate design concepts and proposals. Typically, there are plenty of software capabilities that the organization wants to use, some of which no one knows exactly how to configure (and many times even the consultants).

In defense of consultants, even if he or she has set-up the functionally before, subtle differences in client requirements can affect the configuration settings in many ways. In addition, recalling all the configuration details from previous engagements is not as easy as it sounds.

While configuration classes are a great start and highly recommended, it is unlikely that all possible configuration topics are covered. In addition, while setup checklists or tools are not hard to find, most do not go far enough regarding how to enable or tune the functionality to support different business requirements.

To fill these gaps, *white papers* are the application team's best friend. These white papers should not be confused with sales and marketing white papers—the glossies. Real white papers get into the step-by-step instructions of how to configure the software for specific business scenarios and requirements. The alternative is to reverse-engineer the system by testing different software options, wasting time and money.

The truth is most consultants use white papers. If your consultant does not have the white papers needed, some other resource within the firm probably does. However, it is unlikely you will receive this level of system documentation unless you ask. Consultants do not just hand this stuff to their clients. If they did, the client may no longer need their services!

Beyond your consultants, there are typically other resources to obtain white papers or similar information for your package. These include the software vendor's website or support desk, information sharing communities, books, and from third party vendors providing package specific information (e.g., JDETips.com, ERPTips.com). In any case, start accumulating white papers for each software module and establish a library. Eventually you will need them.

Roles for Software Demos and Design Reviews

Software demonstrations and design reviews are tools to share information with stakeholders, obtain feedback, or validate proposed directions. For the project manager, these activities also represent another knowledge transfer reality check.

When the outside consultants are responsible for performing software demos and leading design reviews, the project team is off the hook. When employees on the team understand that these are their responsibilities, this drives the perceived need to learn, gets them moving out of their comfort zones sooner, and eventually raises their confidence level with the software.

This is not about throwing the team to the wolves. Outside consultant are involved all along, have a support role in these meetings, and step up when necessary to fill the knowledge gaps. Finally, software knowledge does not occur overnight, but progress should be noticeable with each software demonstration and design review.

Construction Phase

Software Setup Revisited

With completion of the design phase, it is time to revisit and refine the software configuration to ensure that it reflects the entire design and to prepare for testing (the first round of conference room pilots).

During the construction phase, review all configuration parameters and processing options again to make adjustments, create new program versions, and round out any fringe areas not yet configured.

Similar to the rough-cut setup for prototyping, the employees on the application team should make the configuration changes. Again, this occurs with the help of the consultant, but this time it should go smoother since, by now, the team should have a much better understanding of the system.

Who Should Document the Software Configuration

Every organization has more or less a unique software setup, so documenting it is critical. Of course, the software configuration will change during testing, but start this documentation at the end of the construction phase.

The team prepares this documentation and the consultant reviews it for accuracy. This is an opportunity to clear up any confusion regarding what each configuration setting enables within the system.

Equally important to include in this documentation are the business reasons the configuration options were selected (versus other options). If the team cannot explain why each parameter was selected, then they really do not understand what it does or how it applies within the business! Also, without this information, eventually someone will make an ill-advised setup change causing issues throughout the system. ERP software is integrated and even simple changes can have a major ripple effect.

Testing Phase

Ownership of Software Testing

Most consultants agree their client should perform the great majority of testing. But when the other knowledge transfer activities have not occurred prior to this time, testing takes much longer, can be very frustrating for the team, and may not be adequate.

Software testing is necessary for many reasons, but in terms of knowledge transfer, there is no better way to learn the system then when it is not working as envisioned and the team must dig deeper to find out why.

Consultants should assist with developing the overall test plan and initial test cases for the first round of testing, but thereafter, the client should write all test scenarios and perform all testing.

Reenter the Power Users

Power users should also participate in conference room pilot testing. When they do, their software knowledge is elevated close to that of the functional analyst, from at least a transaction-processing standpoint.

The final step of user testing is *acceptance testing*. This includes other end-users, some of which have yet to see the new software at this level of detail. This allows for a fresh set of eyes to validate and test the system. In addition, acceptance testing exposes even more users to the system prior to training the masses.

The Other Value of a Parallel Pilot

Running a limited parallel pilot with the current system prior to go-live is an additional type of test, but it also pushes the team's knowledge of the software to the next level (another cycle of learning).

Cutover Phase

The Client Writes the Work Procedures

Once testing is complete, work procedures are finalized (i.e., operating procedures, system manuals, or desk level instructions). These describe how to navigate the system and use it to perform specific job functions within the company at the transaction level (screens and reports). The functional analyst on each application team should write the work procedures. At this point, if they cannot write the procedures, they do not understand the system enough to train end-users.

Power Users Test Drive the Work Procedures

Theoretically, work procedures should be so clear and concise that anyone can pick them up and perform the job using the system (with no assistance). Power users are good resources to take the procedures for a test drive in the system to identify errors and omissions. Once they do, power users are now in a position to assist with end-user training and post go-live support.

Consultants Should Not Train the End-Users

Delivering the end-user training (including preparation of all training materials, data, and teaching the class) is the not responsibility of the consultants. The functional analyst is the best choice to conduct end-user training since this person is a user, understands the new processes, and has been involved with the software details from the beginning. The rest of the application team and power users assist the trainer and provide support to those in the class as necessary.

Due to the need to train end-users just prior to cutover, the training schedules will probably be tight. When it is necessary to run some classes for the same topics concurrently, the next best choice for the instructor is the in-house application consultant (if you have one). The third choice is a power user.

The "Train the Trainer" Pitfall

Be very careful with the concept of "train the trainer" particularly when the person expected to perform the training has not been involved with the project details. It is not realistic to assume someone can be trained several weeks before the go-live and expect him or her to deliver quality training. Too much can be lost in translation from the team to the designated trainer, thus the trainer cannot answer (or correctly answer) questions during the class that are not scripted.

Seven More End-User Training Mistakes

In addition to training too early or having the wrong person teaching the class, listed below are the seven other most common end-user training pitfalls and solutions.

1. **Employees fail to attend.**

    Once the team develops the training schedule, the executive sponsor (not the project manager or the team) should notify all attendees.

The notification includes each employee's immediate supervisor. This sends the message that training is important and attendance is expected. The schedule should be communicated well in advance to avoid vacation or other schedule conflicts.

2. **Ignoring the need for basic computer skills training.**
Yes, today there are still people that require basic computer skills training such as how to use a keyboard, touch-screen or mouse. Those in this position should be trained in basic computer skills before end-user training on the application because those lacking the skills will struggle and slow the rest of the class.

3. **Poorly planned courses.**
A thoughtful agenda, lesson plan, and supporting training materials are critical. Otherwise, there is a higher probability the course will get off track and not address all important topics.

4. **More than transaction processing.**
A common mistake is to jump right into transaction processing (the hands-on portion) at the beginning of the class. Employees first need a broader understanding of the processes and system in order to separate the forest from the trees. The first step is to convey how the system fits within the context of the organization, departments, and remaining legacy systems (the big picture). Next, quickly review the new business process maps that are applicable, while focusing on *what will change* with the new system. These are the major changes affecting how employees will do their jobs. Third, perform a brief demonstration in the system of each training exercise before the students do the tasks on their own.

5. **No hands-on training.**
Training is not just a software demonstration or a PowerPoint presentation. The majority of the course should be hands-on training, first working with system navigation and then using training exercises. The training exercises include the transactions

and data the students will enter into the system to complete each step. Exercises are an abbreviated version of the work procedures. Review and distribute the work procedures at the end of the class.

6. **The training system is not ready.**

   As much as possible, the training system should replicate what the new system will look like when in production (such as menus, hardware, etc.). Also, make sure the pre-loaded training data supports the hands-on portion of the class (the training exercises). Finally, test the security, hardware, and exercises prior to the class. This provides time to identify and fix issues in the training system to avoid class delays and disruptions.

7. **Too many students in a single class.**

   When a trainer has no assistance in conducting the class, limit the class size to eight people performing the hands-on work. If there is assistance, twelve students are about the maximum class size for any topic.

Post Go-Live

The Frontline of Support

With any system implementation there should be a plan to support the users immediately after cutover. Having each application team and power users field user questions and issues is the final phase of learning within the context of the project. Again, outside consultants can backup the team if needed.

End-User Follow-up Training

In spite of the best end-user training, there will be plenty of questions once the system is live. One reason is that some users will not get serious about learning the system until they have to use it to perform their jobs.

Instead of indefinitely addressing user questions on an individual basis, repeating the same answer to many different users, or users getting different answers from different support people, it is time to regroup. Conduct some brief follow-up training sessions after several weeks into the cutover. Develop a list of answers to commonly asked question and solicit new questions prior to these sessions.

Maintaining the Knowledge

The system documentation and the work procedures should be updated on a regular basis after go-live. Also, resources should be in place to train new employees and as employees change jobs within the organization. As mentioned before, without this support, knowledge of the system and how to properly use it will slowly erode.

Figure 10 – Knowledge Transfer Activities by Phase

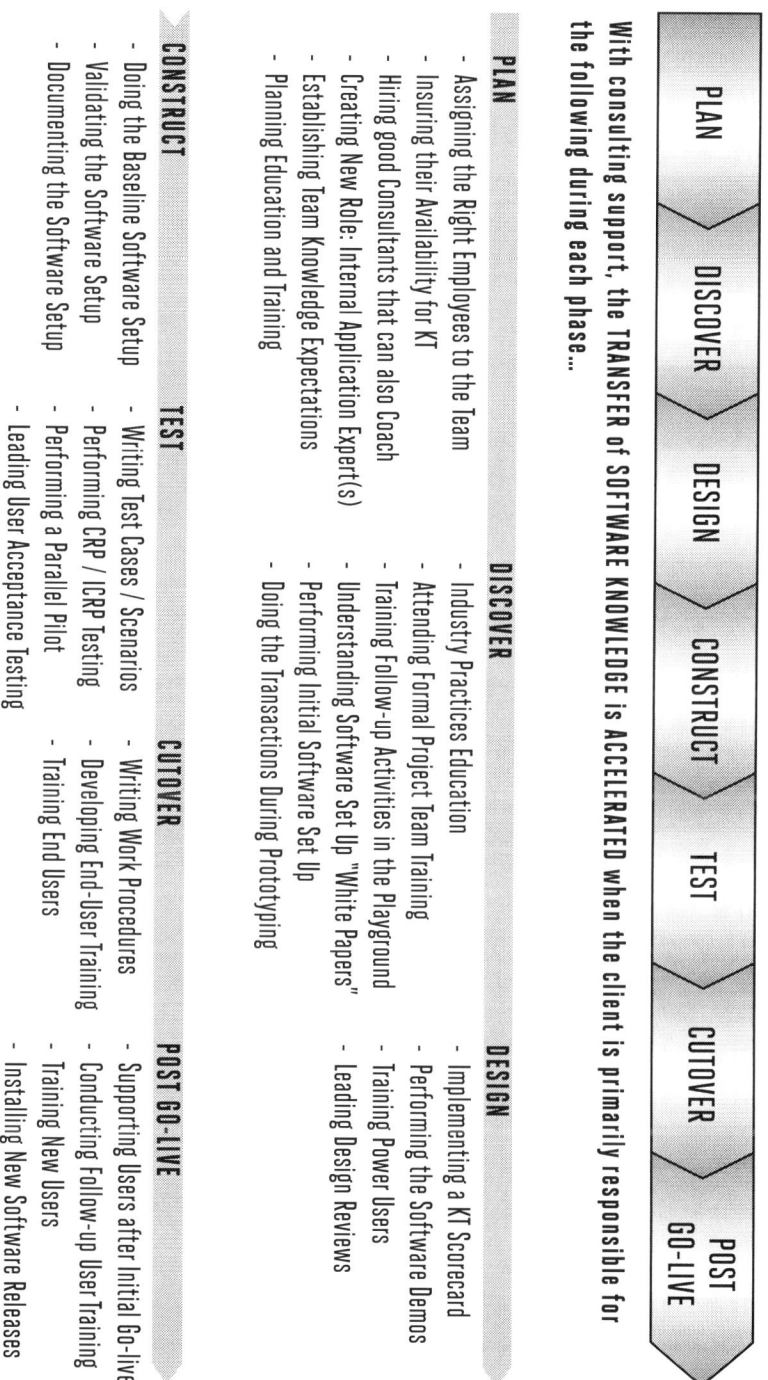

PLAN ▷ DISCOVER ▷ DESIGN ▷ CONSTRUCT ▷ TEST ▷ CUTOVER ▷ POST GO-LIVE

With consulting support, the TRANSFER of SOFTWARE KNOWLEDGE is ACCELERATED when the client is primarily responsible for the following during each phase...

PLAN

- Assigning the Right Employees to the Team
- Insuring their Availability for KT
- Hiring good Consultants that can also Coach
- Creating New Role: Internal Application Expert(s)
- Establishing Team Knowledge Expectations
- Planning Education and Training

DISCOVER

- Industry Practices Education
- Attending Formal Project Team Training
- Training Follow-up Activities in the Playground
- Understanding Software Set Up "White Papers"
- Performing Initial Software Set Up
- Doing the Transactions During Prototyping

DESIGN

- Implementing a KT Scorecard
- Performing the Software Demos
- Training Power Users
- Leading Design Reviews

CONSTRUCT

- Doing the Baseline Software Setup
- Validating the Software Setup
- Documenting the Software Setup

TEST

- Writing Test Cases / Scenarios
- Performing CRP / ICRP Testing
- Performing a Parallel Pilot
- Leading User Acceptance Testing

CUTOVER

- Writing Work Procedures
- Developing End-User Training
- Training End Users

POST GO-LIVE

- Supporting Users after Initial Go-live
- Conducting Follow-up User Training
- Training New Users
- Installing New Software Releases

CHAPTER 17
BREAKING DOWN THE SILOS

A Solution Looking for a Problem?

One of the biggest mistakes during ERP projects is not taking the time to build a common understanding of how business is conducted today and potential improvement opportunities (before design of the new system). Some view this as a luxury, but it is a necessity when more than just incremental improvements are expected from the project.

Usually the thinking is "Let's get on with it. Why bother to analyze our current processes and issues? It is all going to change with the new software. Also, the consultants say they have tools and industry templates to configure the software to meet our requirements. It should be that easy, right?"

Well, things will change for sure. But without a common understanding of current processes, how will we agree upon *what* must change? Second, without a shared definition of the problems and opportunities, how can we design the right solutions or know what potential solutions are even relevant?

Of course, ERP software can enable many improvements, but software alone does not necessarily result in best practices or even better practices than you currently have. Contrary to the general view of technology, ERP software is not plug-and-play, such as using Microsoft Office, mobile apps, or ordering a product on the internet. With these applications, there are no best practices other than pushing the right button! Successfully using ERP to run an entire business is not quite that easy.

Some ERP packages consist of well over a thousand programs, many of which are highly integrated. It is best if these applications are utilized in conjunction with redesigned workflows such as new polices, procedures, roles, and measurements. This makes all the difference in the world, because without these, you are simply rolling the dice. Remember, the organization will fit (or force) itself into the software, one way or another.

Understanding Fragmentation

I must admit, any statement implying that organizations do not understand their business processes sounds a bit ridiculous at first. For this reason, many consider the As-Is analysis to be an unnecessary exercise, or gloss over it.

Collectively employees do understand the processes, but individually they *do not*. That is, within each department, the managers and employees are very familiar with their slice of the world, but their perspectives are somewhat limited. Major business processes cross department boundaries, and this gets in the way of a common understanding and creates barriers to improvement.

Take the Order Fulfillment process as an example. Within this process, the sales department, customer service, warehouse, accounts receivable, and invoice deduction departments are all involved. It is highly unlikely that any single employee has a grasp of the entire workflow, let alone an understanding of the cause and effect relationships of process related issues.

Furthermore, each department is responsible for only a *piece* of the overall process. This results in a narrow focus of concern, conflicting departmental objectives, and a *fragmented* organization. All of this renders managers within each department powerless to affect systemic change; hence, the "silo effect."

In the example above, what authority does the warehouse manager (who is responsible for accurate and timely shipments) have in ensuring that the customer service department get the sales orders entered on time, with the right *ship to* address, the right ship dates, the right carrier, and the correct freight charges?

Therefore, who is responsible for the Order Fulfillment process? Chances are many managers are involved, but no one is empowered to make the holistic changes necessary. Probably many of these changes are obvious to most employees that must deal with the issues every day.

However, most employees would rather not rock the boat, even most of the good ones. Instead, they go about their daily functions and work around the *symptoms* of broken processes. They do what is best for their own department, improve what they can control, or enter a state of denial. Somehow they get the job done, but at what cost? This is not about bad employees, but rather bad business processes.

Over time, "history" shapes the work environment, and sub-optimization becomes a way of life—the *only* way. In fact, new employees are trained to sub-optimize since workarounds are now part of the *standard operating procedures*. This reinforces broken processes. No wonder change is so difficult!

The *As-Is* Opportunity

We all know that business process changes are necessary to take advantage of the new ERP software. However, considering the points made above, it is apparent that many potential improvements have nothing to do with software. Many fail to realize that improvements unrelated to the software can yield the most benefits.

Defining the current processes and issues is the first step in developing better business solutions. The opportunity comes from this: Many processes were never really designed, but evolved to workaround the causes of process deficiencies.

It is likely that no cross-functional team has taken a hard look at the business processes for a long time. When this is the case, usually the improvement opportunities are abundant, easily identified, and easily implemented.

When performing this analysis for an ERP implementation, not every opportunity will be a major improvement. However, it is the accumulation of many improvements across the enterprise that can noticeably increase productivity and reduce the cost of doing business.

Finally, the cross-functional approach is an excellent way to get end-users and other stakeholders (not on the ERP team) involved with the project from the start. This creates more ownership in the eventual solutions.

Are Consultants Required for the *As-Is* Analysis?

Save yourself some money. First, your software consultants should not perform this analysis, especially with the usual approach of separate interviews with different employees.

Furthermore, it is not necessary for your consultants to sit through current process analysis meetings. The company can conduct these meetings and do the analysis on its own with a cross-functional team, an unbiased meeting facilitator, and a deliverable format that is easily understood (discussed later).

As mentioned many times before, a cross-functional team collectively understands the current processes and issues better than any consultant. Also, knowledge of the new system is not required to document the *As-Is*. Often this is a point of confusion. The *As-Is* analysis provides *input* into the design phase. During design, it is determined how the software will be used along with changes in business practices. This is when knowledge of the new software is necessary.

The key is once the current processes, issues and opportunities are documented, a good software consultant can review the deliverables, ask questions for clarification, and provide feedback based on their experience. This can occur in a matter of hours with the team and does not require the consultant to sit through days or several weeks of process analysis meetings.

Finally, the facilitator does not have to understand a great deal about the current systems, but understanding the deliverable and the ability to manage group dynamics in the sessions are important. In addition, the ability to think in terms of business processes is a very helpful skill.

As-Is Analysis Deliverables

The deliverables of this phase come from continuous improvement disciplines. After all, the goal of any ERP project should be to improve the processes. I recommend three deliverables to get a clear understanding of the business operation. These include the *process map, problem analysis*, and *suggested improvement opportunities*.

The process map should capture the workflow and its key attributes. It is constructed during the initial meeting for each process within scope. Do not confuse this with a data or system diagram. The process map not only addresses the information flow, but also each activity performed, the role that is responsible, the systems that are used, and the business rules governing how the work is completed. A process map should represent both the physical and logical aspects of the process.

The process mapping deliverable I prefer is the IDEF format. It has been around for a while and is excellent for three reasons: The format is simple; it captures all key elements of the process listed below (some of which are omitted in other models); and when complete, it communicates easily. For each activity within a workflow, the process map should capture the following:

- **Inputs.** The information or materials consumed within the activity (outputs of the previous activities).
- **Supports.** The entities that perform the activity or support it. These

include a person (job title, not employee names), systems (applications, spreadsheets, etc.), or equipment (e.g., a forklift, a conveyor, a scale).

- **Controls.** The key management policies or rules that should be followed by those performing the activity.
- **Outputs.** The information, document, or product produced as the result of the activity (inputs into the next activity, decision, or wait-time).
- **Wait-Time.** Queue time or delays before the start of the next activity (the "in-basket" or waiting for system reports, etc.).
- **Decisions/Inspections.** The important decision points that determine the next activity to perform.

The problem analysis is completed in the second team meeting for each process and *brainstorming* is the method to gather the information. The deliverable consist of three parts including a list of process related problems, root causes, and the effect each problem has on the business.

The definition of a problem is an issue, barrier, or inhibitor that negatively impacts business performance. After identifying the problems, try to get at root causes as much as possible.

Understanding the *effect* of a problem is an excellent way to better quantify the problem and understand the benefits of addressing it. An effect is usually a substitute process that exists to work around issues and has financial or non-financial consequences.

The final deliverable of the *As-Is* analysis is to identify improvement opportunities. This focuses employees on how to potentially fix the problems. Remember, this is a brainstorming session, so the goal is not to design or develop final solutions. The objective is to get ideas flowing. At this stage, there are no right or wrong answers.

As-Is Analysis Pitfalls

The following are some common mistakes when developing the *As-Is* and how to avoid each:

- **Dusting off old process maps.**
  In the past, someone probably created process maps for some other project. This information may be useful to an extent, but process

analysis is not a presentation; nor do we care about how the processes functioned years ago. Most likely, the old maps are incomplete, wrong, or lack the relevant information for each activity within the processes.

Above all, the value of the *As-Is* analysis is in the cross-functional interactions, discovery, and *shared* learning of those involved. The desired level of participation, consensus building, and ownership is not possible by just reviewing old process maps.

- **Maps are too high-level.**
  When planning the project scope, major business processes were identified and broken down into sub-processes. For example, receiving a purchase order is a sub-process of procurement within any company. It does not hurt to first map at the major business processes level, but to have meaning, the process maps must be created at the sub-process level.

  Next, determining the level of detail to map each sub-process is not a scientific decision. It comes naturally once the meeting begins. The right level is how employees explain the specific steps they perform when doing their job. In addition, it is the facilitator's role to ask questions to capture the activities within each process and associated information.

- **Separate Meetings with Each Department.**
  Conducting separate meetings with each department associated with a process defeats the purpose of the analysis. The goal is to build a common understanding of the workflow and issues, not to emulate the departmental silos.

  Each meeting should include representatives from all departments directly or indirectly involved with the process. Those directly involved include key employees from all departments performing the work. Those indirectly associated with the process provide the major inputs (internal suppliers) or receive the major outputs (internal customers). Depending on the process, external suppliers and customers participating in these meetings may be appropriate. The concept is to get stakeholders in the same room to hash out differences in perceptions. Again, this creates alignment and a more accurate picture of what is happening today and enables a better understanding of how issues affect the business.

- **Excluding the "Doers."**
  This pitfall could be a simple oversight when planning the meetings or a cultural norm that says managers know everything, and those that do the work know nothing. Failure to involve the doers always results in inaccuracies, lack of detail, or sometimes managers portraying a process the way they want it portrayed.

- **No Meeting Ground Rules.**
  In order to get the most out of the meetings, we need the right forum. In addition to a meeting facilitator and a brainstorming format, ground rules are necessary. Enforcing the ground rules is usually the biggest challenge for the facilitator, and the reason you need an unbiased person in this role.

The most important ground rule is that everyone's opinions count. Therefore, managers in attendance must take their stripes off at the door. The idea is to get all business issues on the table—the good, bad, and the ugly. Everything about the process is fair game, and no business issue or opportunity is taboo. The meeting is not about defending the sacred cows or protecting ones turf.

The other important ground rule is the meetings are not a witch-hunt or about blaming people. Broken processes can make good employees look bad at times. The focus should be on the business (the process, issues, and opportunities), not personal attacks. Therefore, all issues should be expressed within the context of the workflow and the effect on business performance.

- **Failure to Document How the Process Really Works.**

  The goal of the *As-Is* analysis is to understand how a business process *actually* works, not how it is supposed to work. In order to identify improvements, there is no value in documenting the "ideal" version of the current process or the standard operating procedures. In most cases, this is not what really happens.

  For example, rework loops, redundancies, and workarounds are part of the *normal* workflow. These are symptoms, and we must acknowledge the symptoms in the process maps to get at the underlying issues.

- **The Analysis Takes Forever.**

  For a typical ERP project, there might be a dozen business processes to map. In order to avoid paralysis through analysis, focus the effort on the mainline processes or combine similar ones into a single session when there are only slight variations in how the work is performed.

  In addition, instead of stringing out the three deliverables (maps, problems, and improvements) over a series of one-hour meetings conducted over many weeks, compress the timeline. For example, schedule several meetings lasting four to five hours, over a period of a week. This increases the focus and conveys a sense of urgency to complete the task. Finally, schedule unrelated process sessions concurrently, but the limiting factor might be the number of qualified facilitators.

- **Senior Management Is in the Dark.**

  The executive sponsor should lay the groundwork for successful As-Is meetings by communicating to the organization the purpose of the analysis. When the analysis is complete, the project team should present to the steering team a summary of the key issues and improvement opportunities identified. This keeps executives involved and provides feedback to the team prior to prototyping the new software or designing new processes.

The *To-Be* Process Design

Some companies have a well-defined vision of how they want to do business in the future before buying ERP software. This vision of future business processes drives the software selection, *To-Be* design, and the software configuration. Of course, this is the ideal situation.

When embarking on an ERP project, there always should be a clear set of objectives established by senior management. However, at the start of the project most companies have only a vague understanding of how future processes should function to support the objectives. I always encourage senior management and other key managers to do more work in defining the *future state* at the beginning of the project. But when management cannot take this vision as far as we would like at this stage, this does not mean the project is destined for failure.

Defining the future state is a journey that should come to fruition when designing the *To-Be* processes. If the company selects a good software package, hires qualified software consultants, properly trains the team, and has an understanding of current processes and opportunities, defining new business processes is not as difficult as it sounds. When this is the case, once into the design phase the solutions are more or less already there. Now it is a matter of putting it all together, filling in the blanks, and further testing.

The *To-Be* deliverables consists of the *gap analysis* and *new process maps*. The gap analysis is important since it identifies differences in what we want to do in the future versus what the software will allow us to do. The application consultant (whether internal or from the outside) plays a major role in this analysis since this person is the software expert. The deliverable depicts the steps within each proposed process design including a description of any software gaps identified for each step.

The goal is not only to identify the gaps, but also to reassess the validity of the items considered to be issues or develop solutions to close the gaps. The solutions may include alternative ways to configure the software, policy and procedure changes, workarounds, or software modifications. Of course, the solutions to fill the gaps will affect the final *To-Be* process design.

The *To-Be* process maps utilize the same deliverable format as the *As-Is*. The major difference is how it is developed. Each application team drives the future process designs and develops the initial proposals. Next, major stakeholders outside the application team (power users, managers, etc.) are called in to provide input and validation at the right times. Note that this should not be the first time they have seen the software. Before this time there should be software demonstrations, knowledge transfer sessions, and design discussions with key stakeholders.

Once initially complete, the new process maps will continue to evolve during the construction and testing phases. The hope is the net result of these activities makes the design better.

CHAPTER 18
TESTING

Have you ever wondered why a software program blows-up the first time it is used, even though those who developed the program insist they previously tested it? Multiply this situation by hundreds of ERP programs that were "tested," and then imagine the chaos at go-live. This scenario is not unusual.

In fact, inadequate testing is a double whammy. It results in many lingering software bugs that are eventually discover once the system is in production. It also leads to less software knowledge for the project team since testing is a significant part of learning. A team that does not understand the software means inadequate documentation of the new work procedures and poorly trained end-users. Once the system is live, lack of user knowledge creates as much confusion within the organization as software bugs.

Much has been written about software quality assurance (SQA) over the years. The purpose of this chapter is to draw upon some best practices that can be applied without the need to become a software quality expert.

Testing Phases
The project should include several types of testing such as the Conference Room Pilot (CRP), Integrated Conference Room Pilot (ICRP), Limited Parallel Pilot, User Acceptance Test, Volume/Stress Test, and a System Cutover Test. Within the CRP and ICRP test phases, there are several *rounds* of testing in each. Each round represents a greater scope and depth of testing.

Conference Room Pilot
The Conference Room Pilot is the first phase of formal testing. It is mainly concerned with testing the programs *within* each software module. As with any testing, this includes the standard programs that come with the package and all custom developed programs.

Each application team is responsible for testing their software module during CRP. The concept is to first test each module of the system independently to ensure it is working properly before adding more complexity.

Integrated Conference Room Pilot

The Integrated Conference Room Pilot is when it all comes together for the first time in the system. The goal is to test all software modules functioning together. Just because individual modules work fine when tested individually during CRP, does not mean they work well when interacting with one another.

During ICRP, the team tests related business processes across multiple modules. Therefore, ICRP requires a higher level of coordination and communication between all application teams when planning and executing the test and when resolving issues.

Back to our simple business process example: ICRP testing of the *Quote-to-Cash* process might begin with a single order transacted through its entire cycle. The processing starts with setting up a customer and items. In general, the next steps are to enter a sales quote, convert it to a sales order, check credit, generate a warehouse request, pick and ship the product, and then invoice the customer. Along the way, there will be behind the scene system updates to the inventory, accounting, and perhaps other modules. Finally, the payment is collected from the customer, invoice deductions are applied, and credits are issued within the accounts receivable module. This also includes more updates to the accounting module.

As one can see, the complete cycle for a major business process may include up to five or more software modules. If something is not setup correctly in one area of the system, the entire process can come to a halt. This is the reason integration testing is very important.

Test Cases

Within most business processes, there are many scenarios or variations to what can occur. In the sales order example previously discussed, there will be order cancellations, backorders, and pricing issues, to name a few. In fact, a single process may have well over a dozen scenarios that occur on a daily basis, let alone, other things that happen less frequently. Moreover, the business scenarios are not mutually exclusive. Many can occur at the same time, and unique customer requirements always add a different twist.

A test case is the vehicle to define each variation to a process and to ensure it is properly tested. It documents a business procedure (scenarios) to be tested, responsibility for testing it, and provides a place to record test results, a disposition, and corrective actions. The first objective is to define specifically what to test. Randomly meandering through the software is not really testing.

A test case that is exercised in the system should have only one of two possible dispositions: Either it *passed*, or it *failed*. In order to pass, the process should proceed through the system with no glitches. Therefore, any software bug or setup issue results in a failed test case. In this situation, the fix is eventually made to the system and the entire test case should be performed again. If necessary, this cycle should continue until the test case passes. It is important to minimize the number of failed test cases that must be revisited in the next round of testing.

In the situation where there are no software issues, but policy or procedural issues still remain, pass the test case, and add the item to the issue list. The actions necessary to resolve the issue could require additional test cases to ensure the solution is working as desired.

Testing Oversights

Listed below are common pitfalls during CRP and ICRP testing:

- **Failure to Unit Test Custom Programs in Advance.**
  Many of the standard programs within the package have been *unit tested* prior to CRP as part of prototyping and the design phase. However, this is usually not the case for programs containing custom code. Unit testing of custom or modified programs should occur prior to CRP (or ICRP) as part of developing the software. This testing should involve the developer, application consultant, and the functional analyst. There is no point in introducing a custom program into formal testing that does not meet basic user requirements or is full of bugs. This slows down CRP or ICRP because of software rework that could have been easily avoided with proper unit testing.

- **Inadequate Test Coverage.**
  A term often used within software quality assurance is *test coverage*. The proper coverage is when the great majority of business scenarios have been tested in the software. While it is not possible

to identify and test every conceivable activity within the system, aim for 110% coverage with the hope that 95% of the scenarios that could occur are tested.

Without this philosophy, the many one-off software bugs and procedural issues from scenarios not tested can accumulate into one major go-live mess. This gets back to defining the test cases. A failure to test a range of business scenarios or deviations to the normal workflow results in a lack of test coverage.

The final point is that users can (and will) make mistakes when using the system. Many of these mistakes are predictable, so develop test cases to create the data issue, and then try to reverse it in the system during the test. The steps to correct common user mistakes should be part of the work procedures and end-user training. This will reduce the number of questions or crises after go-live since users will have some understanding of how to address the problem themselves.

- **Allowing the Consultants to Do Most of the Testing.**
  Each application team should have a functional analyst and team leader. These employees should be from the user area and be knowledgeable about the processes in question. For this reason and for the purposes of further knowledge transfer, they should develop the great majority of test cases and perform 90% of all testing. By now, they should understand the software well enough to test it.

  Software consultants can help develop the test cases for the first round of CRP only (to help jump start the process) and assist with some early testing. All consultants should have some basic scenarios to test first. This allows for getting past the most obvious software issues so the team can take over from there.

  During the time the functional analyst and team lead are testing, the consultant should provide support when needed and periodically verify what is happening with the data in the system. "Black box" testing by the users does not always uncover potential data anomalies lurking in the background that can cause problems later.

- **The Test System Does Not Represent Production.**
  In order to have confidence in the test results, the test environment should reflect what the system will look like when in production. The problem is that many times certain hardware, equipment, security, and menus that will exist in the production system are rolled into testing too late in the game. Therefore, these items receive very limited testing or cause issues that require the team to retest areas of the software that previously functioned correctly. Start testing all hardware and equipment representative of the production environment during CRP. Security and menu setup can wait until the latter stages of ICRP.

- **Using Data Conversion Programs Too Early.**
  It is always recommended to start on the design and development of data conversion programs as early as possible since many are used fairly early in the testing cycle.

  When to introduce data conversion into the CRP is a different topic. Instead of using the data conversion programs for the first round of CRP, load the test data manually for three reasons: First, one does need much data for round one of CRP. In fact, too much data can get in the way.

  Second, if the data is converted using the programs, and then issues arise during early testing, it is more difficult to discern whether the problem is with the software or a data issue caused by the conversion program.

  Third, the first round of CRP testing will probably necessitate changes to the data conversion programs due to new discoveries about the software. Make the changes and then use the conversion programs starting in the second round of CRP.

- **Not Fully Testing Interfaces and Software Modifications.**
  Interfaces and software modifications are usually the most difficult custom programs to write. If not carefully tested, these programs can become the Achilles' Heal of the project. During CRP and ICRP, one cannot test interfaces and modifications enough.

- **Failure to Regression Test.**
  After each round of testing, it will be necessary to make software configuration changes to address problems. Once these changes are incorporated, it is a common mistake to assume other related programs will continue to function correctly. Due to the integration of the system, sometimes a configuration change in one area breaks other areas of the software.

  After verifying an issue is fixed, it is important to retest other programs that could be affected by the configuration change (i.e., regress). Regression testing does not require a new test case and does not have to be extreme; otherwise testing would be a never-ending loop. Nevertheless, some retesting is always recommended.

User Acceptance Test

At some point, the power users and major stakeholders should participate in the CRP and ICRP. But once this testing is complete, they should be responsible for organizing more end-users to conduct a separate user acceptance test.

The concept behind user acceptance testing is that the project team has finished their testing and believes the system is now functioning as required. However, other managers and end-users not involved with the project to this point, may have major concerns.

Acceptance testing represents the last chance for users to identify any major issues prior to system go-live. Though not considered training, acceptance testing is also another opportunity to transfer software knowledge to the user community prior to end-user training.

Acceptance testing is usually less structured than CRP and ICRP. Nevertheless, when this testing is successfully completed, some project managers require a sign-off by managers participating stating the system satisfies user requirements. The nature of the project and company culture determines the necessity for a user sign-off. But when sign-off is required, the users will probably take this testing more seriously.

Limited Parallel Pilot Test

This type of pilot is not performed in a conference room, but within the actual work area. It is limited in the sense that it occurs within a contained area of the business, and it lasts only a few days or perhaps a week. It is parallel since it involves end-users using both the new and current system simultaneously.

No parallel pilot is a substitute for conference room pilot testing, but it provides the important opportunity to use the new software, procedures, interfaces, etc., in the real world.

There comes a point when the team can continue to test in a conference room, but not learn what can be learned in a three-day pilot. This is because it is not possible to anticipate every scenario that can occur in the actual work environment. Not only is it another chance to identify and fix issues that might otherwise fall into the cracks, but it is a great learning experience for the team and users.

It is important to select a pilot location that best represents the scope of the software footprint because pilots take time to plan and execute. An easy or narrowly defined pilot may limit its value, considering the effort required to conduct a pilot.

For example, the product service department is often an excellent pilot location because of the software coverage. The department is somewhat contained but may use most of the software functionality to be used by other departments. These processes might include sales orders, pricing, purchasing, inventory, and work orders.

Volume/Stress Test

When system performance or stability issues arise at any time (even subtle ones), do not ignore them. Beware if the system is slower or crashes occasionally during testing or end-user training. You had better find out why and address the problem, because there will be many more users pounding on the system when it is in production.

To be proactive, some level of volume testing is recommended. This type of test is about throwing as much load on the system as possible to be reasonably sure the system can handle the transaction volume expected in the production environment. It is usually not possible to replicate the load that will occur in production. Nevertheless, in the test system it is wise to run batch programs that consume a large amount of system resources all at the same time, while a fair amount of users are performing interactive transactions. If transaction processing is slow during this test, it will probably be a lot worse in production.

System Cutover Test

When it is time to cutover to the new system, usually the business operations must cease temporarily in order to get a clean transition from the old system to the new (though normally some early conversion steps can occur while the business is still running).

Therefore, in order to perform the conversion, there must be a window of time with no users on the legacy system to successfully complete the majority of steps in the cutover plan. This plan could be lengthy and includes some conversion activities that are automated and some that are manual, and must occur in a very precise sequence.

The cutover window is usually an agreed upon weekend or a weekend with a holiday (depending on the time required to convert). When working with management, scheduling this window is not necessarily easy since there is always work to perform within the organization.

Whatever the conversion window happens to be, you do not want to blow it with a cutover that is aborted halfway through because of a major problem. Finding another acceptable cutover window could cost the project weeks or even months. Second, a flawed system cutover could cause a rollback to the old system after only several days running the new system. This is also the reason you should always have a rollback plan.

In order to increase the chances that the conversion will go smoothly, do a dry run test of the conversion plan prior to the real go-live.

CHAPTER 19
STAYING ON TRACK

In many ways, items associated with ERP success—quality software, benefits realization, a project that is on-time and within budget—are simply *by-products* of embracing the philosophies, strategies, and techniques discussed in previous chapters. This is important because controlling the project is much easier when we have done most of the other things correctly.

Project Control Mechanisms

Beyond the most obvious project management practices of tracking actual expenditures versus the budget and progress versus the schedule, there are numerous other tools and techniques to maintain a handle on the project:

The War Room

First, it is important to get team members out of their norm work environment so they can focus on the project. When possible, all project related work takes place in a room dedicated for this purpose. Also, the teams working in close proximity to each other enhances informal communication, which can be the best communication of all.

Project Administration

On a shared network drive, create directories for the team to store all project deliverables and other documentation. I find it best to organize the folders by team and then by project phase. In addition, establish a calendar where all project meetings are recorded and visible in one place.

Time Compression

Delays in making decisions and resolving issues can cost a project dearly, especially if the decision or issue is on the critical path. Much of this has to do with the drawn out nature of meeting schedules. The more the project manager can compress the allowable time to make decisions and resolve issues, the shorter the timeline. *Time compression* creates more focus and a sense of urgency. Of course, setting aggressive dates for decisions and issue resolution is not about forcing solutions, but we must keep the project moving.

There are two ways to accomplish this objective. The first is to increase the duration of meetings. This reduces the number of meetings that might otherwise occur over a longer period.

The second is to have shorter meetings but very frequently. For example, a daily fifteen minute status meeting until a decision is made or the issue resolved. This method is best suited when much of the analysis or work is performed outside this meeting.

Quality Reviews

It was mentioned previously that project deliverables represent the building blocks of the project. When major deliverables are theoretically complete, it is a good practice to have an informal quality review (wrap up session) with all stakeholders present. Because it is considered a final review, there should be few surprises, but usually enough changes are made to make most quality reviews worthwhile. Again, quality reviews do not imply there is separate quality control function, since each team is responsible for the quality and completeness of their work.

Containing Scope

The project manager must be vigilant about adhering to the original project scope, but a procedure is necessary to manage the inevitable requests to expand the scope. A procedure makes the final decisions less subjective and less personal (i.e., a business decision).

The steering team has the responsibility to approve or reject requested changes in scope. However, the project manager should have the latitude to determine when a requested change requires steering team approval. In this case, the project manager can approve or reject the request.

For example, the addition of a few reports or minor screen changes probably does not warrant steering team involvement. Also, sometimes a minor increase in scope accommodates some significant and unexpected project benefits.

A requested change should be documented on a *scope change request* form (SCR). It includes the name of the requestor, describes the requested change, alternatives to making the change, the business benefits of the change (quantified when possible), the estimated impact on the project schedule and budget, and the risk of doing and not doing the change. For example, if the change is for a major software modification, the risk assessment should address how invasive the change is within the software. This is an indication of how difficult it might be to retrofit the mod into a new software release.

Once the SCR form is completed by the requester and the project manager, the requester presents the change request to the steering team for their approval or rejection. That is, if the requester feels strongly enough about the scope increase; he or she should be prepared to defend it (not the project manager).

Managing Issues

Always remember that project issues are "good" in the sense if there are no issues, the project is probably in trouble. However, it is critical to be religious about maintaining the *issue list*. The project manager must make sure that issues are addressed early and often. Unresolved issues can quickly compound, thus, delaying the project or causing rework.

There should be a centralized list of all project issues. The project manager may add to the list, but each team is responsible for adding their issues and maintains the status of each in the list.

In documenting each issue, record the originator, issue date, the type of issue (software bug, procedural, etc.), priority, status (open or closed), to whom it is assigned, scheduled resolution date, the actual resolution, and date resolved. Keep all closed (canceled or resolved) issues in history since sometimes people have amnesia when it comes to agreed upon resolutions.

The Team Task List

The Master Schedule established the target dates for the start and completion of major deliverables. The Detail Schedule includes the steps to achieve each deliverable, with dates, and those responsible. It is a tool to help control the project, and becomes the baseline by which progress is measured.

In spite of all the necessary planning, ERP projects are dynamic in nature. While the project goals and master schedule hopefully remain the same, the detail steps to get there will in fact change. As things change, we need a quick mechanism to reset priorities and tasks in the near-term to keep the teams focused on achieving the overall objectives (without constantly revising the detail schedule). The *task list* is the means to accomplish this.

The task list covers the activities planned for each team over the next three weeks. The project manager establishes the priorities for the period, defines the major tasks to be completed, and sets direction on how to achieve each. Each team is responsible for maintaining its task list. Also, each team will have activities specific to their software modules or things they must accomplish that are not necessarily assigned by the project manager. These tasks should also be included in their task list.

The teams maintaining their task list makes them more responsible for managing their activities. The project manager must be able to provide direction and answer questions regarding assigned tasks, but cannot plan every activity of every team.

Firm Up The Consulting Schedules

During the planning phase, a consulting budget was developed that included the estimated number of hours per week for each consultant. The budget was communicated to the consultants at the beginning of project. This enables them to plan, and allot time for the project, but it does not give them the approval to use the hours exactly as budgeted.

In order to control the actual consulting hours consumed, every week the hours are firmed up by the client project manager to cover the next three weeks for each consultant. Consultants then are expected to work within the firm hours for the week. If more time is required within a given week, the client project manager must approve the change before the consultant can work the additional hours.

This approach helps protect the consulting budget since the firm schedule is based on what is occurring on the project at the moment. This better matches the hours to the consulting needs (versus much earlier projections).

Without this approach, there is a tendency to over-schedule consulting hours, burn them, and then wish you had more hours. Hours are wasted because once a consultant has the hours scheduled, the consultant will use them (perhaps not in the most productive fashion).

Finally, to create a hedge, it does not hurt at times to cut consulting hours slightly below what might be needed. Over time, if a consultant is not receiving the hours necessary to stay on schedule you can increase the hours accordingly.

When using consultants sparingly, send them a list of your questions before they arrive. This helps them prepare so they can be more productive.

Keeping Time

It is important to ensure that the time committed to the project for each team member is occurring according to the plan. The hours spent on the project should be recorded in a centralized spreadsheet or database at the end of each week. It is useful when the hours are reported against the major project deliverables. This reporting also includes consultants.

Project Management Team Meetings

The project management team should meet every two weeks to assess progress, team performance, and layout strategies and plans going forward. The meeting includes the project manager, consulting project manager, IT director, and site manager (associated with the current project phase). The executive sponsor may be asked to attend these meeting when required.

Project Team Meetings

The entire project team (project management, application, and technical teams) should meet every week. Each application and technical team prepares a status report for review during the meeting. The status update includes progress (versus the task list), new issues, and new team concerns. Note: This meeting is not about resolving issues (unless the issue can be addressed quickly and by only the project team). Most meetings to resolve issues should occur separately and include the right people.

The One-On-One Team Meetings

Periodically, the project manager should meet with each application team separately to get a feel for team dynamics, effectiveness, knowledge transfer progress, and to address any issues regarding how the team is functioning.

Executive Steering Team Meetings

The steering team should meet every four weeks. The executive sponsor chairs the meeting, and the project manager develops the agenda and facilitates the discussions.

Since the steering team's job is to help manage the project, the agenda should include a project manager update in all of the major project management threads. These include scope management, resources, knowledge transfer, benefits realization, change management, software quality, schedule, budget, technology, and risk. If there is no news in the thread, just say that there is nothing new.

One of the most important parts of the agenda is a *list of major issues and decisions* that have been elevated to the steering team to address. Items should be on this list for one of two possible reasons. First, only senior management truly has the authority within the organization to resolve the issue or make the decision. Second, in spite of their best efforts, the project team and stakeholders could not come to agreement on how to resolve the issue at their level.

How to Contact the Author

I wish you the best of luck with your ERP project. I hope that the knowledge gained from this book will help your organization take more control of its ERP destiny. But as a back up, I provide education and coaching services specific to the project. I am also available for speaking engagements. I can be reached at *stevphillips1@gmail.com*.

REFERENCES

Beaubouef, G.B., *Maximize Your Investment: 10 Key Strategies for Effective Packaged Software Implementations*. Birmingham, UK: Packt Publishing, 2009

Jarocki, T.L., The Next Evolution – *Enhancing and Unifying Project and Change Management*. Version 2.0. Princeton, NJ: Brown & Williams Publishing, 2011

Koch, C., *CIOs Take Back Control of Enterprise Projects from Consultants.* CIO.com, IDG Communications, 2002

Wallace, T.F., and Kremzar, M.H., *ERP: Making It Happen.* New York, NY: John Wiley & Sons Publishing, 2001

Printed in Great Britain
by Amazon